Biological Individuality

In this book, Jack Wilson brings together two lines of research, theoretical biology and analytic metaphysics, that have dealt with the individuation of living entities in virtual isolation from one another. Wilson presents a new theory of biological individuality that addresses problems that cannot be solved by either field alone. The wide range of unfamiliar and fascinating organisms that he uses to develop his view, including slime molds, parasitic barnacles, and tardigrades, enables him to escape the limitations of theories based on thought experiments and the timeworn examples of organisms on which philosophers have traditionally relied. He presents a more fine-grained vocabulary of individuation based on diverse kinds of living things. This allows him to clarify and resolve previously muddled disputes about individuality in biology and philosophy.

This is a clearly written book of interest to philosophers of biology, metaphysicians, and biologists.

Jack Wilson is Assistant Professor of Philosophy at Washington and Lee University.

CAMBRIDGE STUDIES IN PHILOSOPHY AND BIOLOGY

General Editor
Michael Ruse   *University of Guelph*

Advisory Board
Michael Donoghue   *Harvard University*
Jean Gayon   *University of Paris*
Jonathan Hodge   *University of Leeds*
Jane Maienschein   *Arizona State University*
Jesus Mosterin   *University of Barcelona*
Elliott Sober   *University of Wisconsin*

Alfred I. Tauber: *The Immune Self: Theory or Metaphor?*
Elliott Sober: *From a Biological Point of View*
Robert Brandon: *Concepts and Methods in Evolutionary Biology*
Peter Godfrey-Smith: *Complexity and the Function of Mind in Nature*
William A. Rottschaefer: *The Biology and Psychology of Moral Agency*
Sahotra Sarkar: *Genetics and Reductionism*
Jean Gayon: *Darwinism's Struggle for Survival*
Jane Maienschein and Michael Ruse (eds.): *Biology and the Foundation of Ethics*

Biological Individuality

The Identity and Persistence of Living Entities

JACK WILSON
Washington and Lee University

CAMBRIDGE UNIVERSITY PRESS
Cambridge, New York, Melbourne, Madrid, Cape Town, Singapore, São Paulo

Cambridge University Press
The Edinburgh Building, Cambridge CB2 8RU, UK

Published in the United States of America by Cambridge University Press, New York

www.cambridge.org
Information on this title: www.cambridge.org/9780521624251

© Jack Wilson 1999

This publication is in copyright. Subject to statutory exception
and to the provisions of relevant collective licensing agreements,
no reproduction of any part may take place without the written
permission of Cambridge University Press.

First published 1999
This digitally printed version 2007

A catalogue record for this publication is available from the British Library

Library of Congress Cataloguing in Publication data
Wilson, Jack, 1968–
Biological individuality : the identity and persistence of living
entities / Jack Wilson.
p. cm. – (Cambridge studies in philosophy and biology)
Includes bibliographical references.
ISBN 0-521-62425-8
1. Biology – Philosophy.   2. Individuality.   I. Title.
II. Series.
QH331.W555   1999
570′.1 – dc21   98–32177
                      CIP

ISBN 978-0-521-62425-1 hardback
ISBN 978-0-521-03688-7 paperback

For Marjorie

Contents

Acknowledgments		*page* xi
1	Beyond Horses and Oak Trees: A New Theory of Individuation for Living Entities	1
	1.1 Introduction	1
	1.2 The Meaning of 'a Life'	2
	1.3 The Poverty of Examples	5
	1.4 Imaginary Examples and Conceptual Analysis	9
	1.5 What Is It?	16
2	The Biological and Philosophical Roots of Individuality	22
	2.1 Why Biologists (Should) Care about Individuality	22
	2.2 Philosophers on Living Entities	27
	2.3 Natural Kinds and Substantial Kinds	35
	2.4 Patterns and Natural Kinds	42
3	Individuality and Equivocation	48
	3.1 Paradigm Individuals: The Higher Animals	48
	3.2 Other Possible Solutions	56
	3.3 The Proposed Solution	59
4	The Necessity of Biological Origin and Substantial Kinds	69
	4.1 A Valid Argument for Sortal Essentialism	69
	4.2 The Necessity of Biological Origin	72
	4.3 Sex	80
	4.4 Species Membership and the Necessity of Genealogy	82
5	Generation and Corruption	86
	5.1 Genetic Individuals	86
	5.2 Functional Individuals	89
	5.3 Developmental Individuals	99
	5.4 Raising the Dead	101

6	Personal Identity Naturalized: Our Bodies, Our Selves	105
	6.1 Human Beings as Biological Entities	105
	6.2 Is a Person a Human Being?	111
	6.3 Conclusions	117

Appendix. Identity and Sortals: Why Relative Identity Is Self-Contradictory — 119

Notes — 127

References — 131

Index — 135

Acknowledgments

I owe thanks to many people. The ideas in this book grew out of the work I did as a graduate student in Duke University's Department of Philosophy, which could not have provided me with a more supportive environment. My dissertation committee, Robert Brandon, Michael Ferejohn, Owen Flanagan, and Louise Roth, sent me out the door with excellent suggestions for improving the structure of my arguments. David Sanford, my advisor, lavished supererogatory editorial attention on my work and made the whole process fun.

I am not going to thank the Moortgrat family, the brewers of Duvel Ale, but somehow or other I ended up in contact with Mike Ghiselin after the ISHPSSB meeting in Leuven, Belgium. His help has proven to be invaluable. Over the course of two visits to Mike's house in the California wine country he provided me with great examples and encouraged me to write this book. Once I had, he commented on a complete draft of the manuscript – all that plus grilled salmon and excellent wine. He got me started and Michael Ruse goaded me into finishing it with a series of short, uncensored e-mails. Also at Cambridge University Press I want to thank editor Gwen Seznec and copyeditor Walter Havighurst and an anonymous referee who had many helpful suggestions, particularly on my interpretation of John Locke.

I read parts of this work as papers at Illinois State University, University of Guelph, University of Western Ontario, and Virginia Tech. Special thanks are due in particular to Lisa Gannett, Moira Howe and the other members of B.I.P.E.D., Dick Burian, and Chris Horvath for tough questions. At Northwestern University, David Hull, Arthur Fine, Mathias Frisch, Michelle Little, and Florence Hsia read the penultimate draft and gave me more last-minute advice than I could handle. Daniel Janzen of the University of Pennsylvania and James Anderson of the University of Toronto sent detailed responses to my questions about their work.

Washington and Lee University has generously funded summer research and writing through a pair of Glenn Grants that allowed me to finish the book.

Acknowledgments

Also at Washington and Lee University my colleagues Lad Sessions and John Knox provided thoughtful responses on early drafts and Karen Lyle helped to prepare the final manuscript.

My deepest debt is to Marjorie Schiff. She edited a complete draft of this manuscript and was a constant source of love and encouragement that helped me to see this project through to the end.

1

Beyond Horses and Oak Trees

A New Theory of Individuation for Living Entities

> A main cause of philosophical disease – a one sided diet: one nourishes one's thinking with only one kind of example.
>
> Ludwig Wittgenstein

1.1   INTRODUCTION

Past attempts to explain how to individuate living things have failed for two reasons. They have not assimilated a full range of biological examples or they have been misled by the most common examples and thought experiments. In this book, I explore and resolve paradoxes that arise when one applies past notions of individuality to biological examples beyond the conventional range. I also present a new analysis of identity and persistence.

My argument is based on the belief that to answer the philosophical question "What is a living individual?" it is necessary to find a satisfactory solution to the question "What should a population biologist count when she counts organisms?" Both questions seem to have clear answers when we consider stock examples. Under normal circumstances we can count the number of puppies in a litter or tomato plants in a garden. However, the same intuitions that allow us to count puppies and tomato plants with confidence leave us perplexed when we try to count colonial siphonophores like the Portuguese man-of-war. Things get strange when we extend folk notions of individuality beyond folksy uses. We can find cases in which criteria of individuation for living things that we are used to seeing hang together give contradictory answers to the question "Is it an individual?" If we take the word 'individual' to be synonymous with 'particular,' there will not be many questions at the level of the organism and below (though there may be confusion about the nature of species). But traditionally the term 'individual' has been used more broadly, and in this work I explore many of these uses as they relate to organic organization, genetics, development, and models of natural selection.

The theories of individuation generated by considering only a narrow and conventional range of examples prove inadequate when applied to real living things whose normal modes of existence include complex metamorphoses, regeneration of lost parts, splitting apart and fusing together. A clonal population of the fungus *Armillaria bulbosa* occupies at least fifteen hectares in a Michigan forest. Some mycologists have called it the largest individual living thing on earth. What are the grounds for this claim? Some species of rhizocephalans, a group of parasitic barnacles, have several distinct developmental phases. Is each phase a separate individual or do they collectively compose an individual? Strawberries can reproduce through sexual or clonal reproduction. Is each clone an individual or does the entire set of clones compose an individual? Or are both individuals? Questions like these cannot be answered satisfactorily by a theory that treats the characteristics of a higher animal as the necessary and sufficient conditions of individuality. In fact, cases like these raise the question of whether there are necessary and sufficient conditions for individuality *simpliciter*.

In answering these questions I will address others. What makes a biological entity an individual as opposed to a colony or a component of a larger individual? What criteria should we use to determine that a biological entity – for example, a colony of termites or an asexual organism – is the same colony or organism as one that existed at a previous time? In metaphysical terms, what biological (or other) processes cause substantial change?

In this chapter, I show that past philosophers have failed to explicate the conditions an entity must satisfy to be a living individual. I then explore the reasons for this failure and explain why we should limit ourselves to examples involving real organisms rather than use thought experiments.

1.2 THE MEANING OF 'A LIFE'

Many philosophers assume that it is easy to individuate living things. In this section I present a pair of examples. John Locke claims in the second edition of *An Essay Concerning Human Understanding* that a plant or animal need not be composed of exactly the same particles of matter throughout its existence. A living thing's persistence is not contingent on its particular material constitution. Instead, the continuation of a life preserves the identity of an organism through the flux of material constituents.

> In the state of living Creatures, their Identity depends not on a Mass of the same Particles; but on something else. For in them the variations of great parcels of

1.2 The Meaning of 'a Life'

> Matter alters not the Identity: an Oak growing from a Plant to a great Tree, and then lopp'd, is still the same Oak: And a Colt; grown up to a Horse, sometimes fat, sometimes lean, is all the while the same Horse: though there may be a manifest change of parts. (Book II.xxvii.3)

The metabolic processes involved in the continuation of a particular entity's life result in a constant change of matter. If the continued identity of a living entity does not depend on its being composed of exactly the same matter throughout its existence, on what does it depend?

Locke thinks that the continuation of *a life* preserves the identity of an organism through the flux of material constituents. A plant, for example, persists through changes in its constitutive matter by continuing the same life.

> That being then one Plant, which has such an Organization of Parts in one coherent Body, partaking of one Common Life, it continues to be the same Plant, as long as it partakes of the same Life, though that Life be communicated to new Particles of Matter vitally united to the living Plant, in a like continued Organization, conformable to that sort of Plants. (Book II.xxvii.4)

Similar conditions of identity are true for animals.

> An Animal is a living organized Body; and consequently, the same Animal, as we have observed, is the same continued Life communicated to different Particles of Matter, as they happen successively to be united to that organiz'd living Body. (Book II.xxvii.8)

For Locke, identity is preserved in the changing of substances by being unified by one continued life. A plant that is spatiotemporally connected by a continuous series of matter changes to an earlier plant of the same kind is identical with that earlier plant just in case there is a common life between them. Similarly, the identity through time of a human being or other animal consists in its participation in a common life.[1]

According to Locke, if bodies existing at different times are connected by a common life that endures through the change of material substance, those distinct bodies compose the same living individual at different times.[2] Curiously, Locke does not explain how to individuate *a life*, though this concept provides the principle of individuation for plants and animals. Perhaps he thinks it too obvious to require an explanation – he is, after all, talking about human beings, horses, and oak trees. Under normal circumstances, a competent observer has no trouble determining whether the horse eating an apple today is the same horse he brushed yesterday. Nor does he doubt that the oak tree struck by lightning last winter is the same oak that he carved his initials in as a child.

Three hundred years after Locke wrote the *Essay*, Peter van Inwagen argues that the only way composite objects can exist is if the parts composing the object are connected through a special kind of causal connection; those parts must constitute *a life*. According to van Inwagen,

> ($\exists y$ the xs compose y) if and only if
>
> the activity of the xs constitutes a life (or there is only one of the xs).... I mean the word "life" to denote the individual life of a concrete biological organism. (1990, pp. 82–83)

To phrase the matter in van Inwagen's terms, what is the individual life of a concrete biological organism? He offers some insight into what makes something living rather than nonliving, but does not explain how to decide exactly what constitutes the individual life of an organism. He offers analogies between an organism and a club that is arranged like the metabolism of an organism. He also describes a life as an "unimaginably complex self-maintaining storm of atoms.... One might call it a homeodynamic event." These analogies offer some insight into his intentions, but leave many difficult cases unexamined. He asks again,

> But what is a life? What features distinguish lives from other sorts of events? In the last analysis, it is the business of biology to answer this question, just as it is the business of chemistry to answer the questions 'What is a metal?' and 'What is an acid?', or the business of physics to answer the question 'What is matter?' (1990, p. 84)

Despite his intentions, van Inwagen does not provide a solid set of criteria to distinguish living individuals from parts of larger living individuals or groups of living individuals. I agree with van Inwagen that we will not find the answer to the question of what a life is without reference to "the business of biology," but the biological literature on individuality could itself use some philosophical tidying up.

Determining the boundaries of a life is a more difficult task than looking over the normal range of examples may lead one to believe. Locke and van Inwagen are in good philosophical company when they treat the concept of a life as an intuitively clear idea that can be used to explain other, more difficult concepts, such as identity through time or issues of mereology. But they have not provided a comprehensive description of living individuality. Assuming that we could articulate necessary and sufficient conditions for being alive (and no one has), we still do not know whether a particular mass of living tissue is a living being. It may be, but it could also be several living things or a part of a more comprehensive life. These questions are unanswered by Locke, van Inwagen, and many other philosophers. Why this lacuna? I suspect that both blindspots are, to a large extent, the result of poor examples.

1.3 THE POVERTY OF EXAMPLES

Normal people in normal circumstances can count the number of horses entered in the Kentucky Derby or the number of oaks that flank the driveway. Horses and oak trees are easy to count. They are distinct from their surroundings and other horses or oaks.[3] If we limit our scope to these familiar cases, we can forgive the omission of details as to what is involved in being an individual life. But if we look beyond those cases, we find that our intuitions can lead us to paradoxical judgments of individuality and make us turn a fresh eye to old cases as our basic assumptions about individuality become contingent facts to be explained.

There is a strange poverty of real examples in the philosophical literature on identity. In an unscientific survey of the philosophical literature from Aristotle to the present on the individuation of living things, I have found that the choice of examples breaks down into four basic categories: common plants and metazoan animals, people, artifacts, and science fiction fantasies. The majority of living things, which are neither human nor familiar plants or animals, are absent from this list.[4] The poverty of real examples is matched only by the oddity of the thought experiments involving grossly mutated dogs, human beings who split like amoebae, werewolves, and other products of the imagination.

There is nothing wrong with using horses and oak trees as examples. But these familiar examples are dissimilar from most living things; we cannot limit our examples to a pool of familiar organisms. Horses and oak trees are quite different from each other, but they share many attributes we may not notice at first glance:

a. They are clearly demarcated from their surroundings and other organisms of the same kind. This makes them easy to count.
b. They can reproduce sexually.
c. Each develops from a single cell.
d. Each is (at least mostly) genetically homogeneous.
e. Each is multicellular.

If all or most living things had these characteristics, the same principle that worked for horses and oaks would work for other living things.

Most living things do not share all of these features. A horse is a fine example of a mammal. An oak tree is a fine example of a tree. Most living things, in terms of number of species, number of organisms, or pure biomass, are neither mammals nor trees and do not share properties *a–e*. We ignore the majority of living things if we limit discussion to just these kinds of cases. It is not surprising that our examples would be of this nature. When we think of living things, common, relatively large, and discrete plants and vertebrates are what generally come to mind. It makes sense that our concepts and language have developed around these kinds of living things rather than around colonial invertebrates or giant fungi.

In a five-kingdom taxonomy of life on earth, oaks and horses represent only two kingdoms, the plants and the animals. 'Represent' may be the wrong word. Oaks and horses are only parts of their respective kingdoms, which also contain radically different forms of life. I will discuss these more unusual forms in later chapters, as well as puzzlers from the other kingdoms. How should we count these organisms? Even if we know all the relevant functional and historical facts about the living thing in question, the answer still may not be clear. The suggestion to count individual lives is not much help because that replaces our question with an equally difficult one. Locke's criterion that identity is preserved when there is a common life is little help in deciding these cases because it is not clear what *a life* consists in for that kind of entity. Locke presents an incomplete analysis of what it is to be a persisting individual living thing.

Artifacts have been used as examples at least since Aristotle wrote the *Metaphysics*.[5] Such examples are misleading because of the significant differences between artifacts such as houses, statues, and axes on the one hand, and individual living organisms on the other. A statue of a dog does not, by itself, change its constitutive matter. A dog does. Because of the differences, it is dangerous to attempt to theorize about the identity of living things based on what one believes about the identity of nonliving objects that, if they change their matter at all, do so slowly. Living things are not artifacts. A living thing can change its constitutive matter in a fashion and at a speed that inanimate objects rarely rival.

I turn my attention now to a brief description of a few real cases that are difficult to explain using the commonsense notion of an individual. I will describe aspects of the biology of a colonial siphonophore, a cellular slime mold, and a butterfly. Each presents a unique problem for the commonsense notion of an individual life.

Some colonial invertebrates form colonies that are integrated to the extent that they are functionally indistinguishable from a metazoan individual. The development and behavior of the siphonophores demonstrate the complexity of the problem. A colony of *Nanomia cara*, for example, looks very much like a jellyfish if it is not examined too closely, but it develops by a radically

1.3 The Poverty of Examples

different method. A scyphozoan jellyfish begins life as a single-celled hydrozoan that develops into a multicellular larva. This larva undergoes a series of divisions and ultimately becomes a multicellular body or polyp. The polyp strobilates to form medusae or adult jellyfish.

A colonial siphonophore also begins as a zygote. The zygote divides and forms a larva. The larva's ectoderm thickens and buds off zooids. The process is called astogeny and it is quite different from the development of the true scyphozoan jellyfish. The zooids remain attached together rather than becoming detached. New zooids are budded off from one of the two growth zones located at the end of the nectophore region.

Each colony is composed of a variety of zooids that closely resemble the parts of a normal jellyfish. The top of the colony is a gas-filled float. Below the float are the nectophores that move the colony by pumping water. Their action is coordinated. Other zooids called palpons and gastrozooids ingest prey and distribute the nutrients to other colony members. Sexual medusoids propagate new colonies by forming and fertilizing gametes.

The colony can swim and feed like a single organism. Despite its functional integration, clear vestiges of its colonial nature can be found. Each nectophore has an independent nervous system, but these are coordinated through the nerve tracts connecting the nectophores. The gastrozooids and palpons all pump at the same time (E. O. Wilson 1975).

Both the true jellyfish and the siphonophores have essentially the same functional structure despite their different developmental histories.

> Other higher animal lines originated from the mesoderm, without passing through a colonial stage. The end result is essentially the same: both kinds of organisms escaped from the limitations of the diploblastic (two-layered) body plan and were free to invent large masses of complicated organ systems. But the evolutionary pathways they followed were fundamentally different. (E. O. Wilson 1975, p. 386)

Is a siphonophore colony an individual or is each zooid an individual? Our commonsense notion of individuality does not decide this case, nor does the suggestion to look for the individual life.

The commonsense notion of an organism does not give a clear account of the transition between a caterpillar and a butterfly. A caterpillar develops from a zygote into a complicated multicellular body. Before metamorphosis, it surrounds itself with a cocoon. Inside the cocoon, the caterpillar body breaks down and the dissolved body is used to fuel the growth of the imaginal discs. These discs are small groups of undifferentiated cells that are encapsulated during caterpillar development and play no role in the functioning of the caterpillar's body. When exposed to the right hormone, they grow into the parts of the adult butterfly body and replace the juvenile body. The butterfly is genetically identical with the caterpillar but it is the result of a distinct developmental process fed by the larval body. I am not sure that the commonsense notion of an individual can give us a clear answer as to whether we have one life or two here. Some organisms have life cycles with several developmental sequences as radical as that dividing the caterpillar and the butterfly. To further confuse the issue, there are many degrees of metamorphosis and some stages are composed of more than one organism.

Blackberry plants reproduce both by sexual means resulting in seeds and also through vegetative growth. Some stands of blackberries are hundreds of years old and trace their origin back to a single sexually produced seed. The seed grows into a plant, which sends out runners. Some of the runners and roots remain connected underground and others have become detached. What should we count when we count blackberry plants? The descendants of the initial zygote, the genetically identical descendants of the zygote, or all the contiguous parts of the blackberry plants? Similar problems arise for counting some species of fern, quaking aspen, bamboo, and some fungi.

At one point in the life cycle of certain species of cellular slime molds, a number of independent, ameobalike single cells aggregate together into a grex. The grex is a cylindrical mass of these cells that behaves much like a slug. It has a front and back, responds as a unit to light, and can move as a cohesive body. The cells that compose a grex are not always genetically identical or even related. They begin their lives as free-living single-cell organisms. The grex has some properties of an individual and behaves very much like one. The commonsense notion of individuality does not enable us to determine whether or not it is an individual. These cases break down the connection between the set of properties characteristic of those organisms we feel most comfortable calling individuals using our commonsense notion of individuality.

I will not begin by specifying a set of necessary and sufficient conditions for being an individual living thing. Instead, I will enumerate the characteristics held in common by paradigmatic individuals. I approach the issue this way because I believe that we have developed a rough-and-ready concept of biological individuality as a conceptual tool to help us deal with our practical affairs. There may not even be a complete definition of 'living individual' but we think we know one when we see it. This concept may not prove to be as useful if we try to use it in a philosophical or scientific context. The original concept was not formulated to deal with these apparently paradoxical cases. I want to offer a useful extension of our practice. This extension is not implicit in the rules set by the concept as it currently exists in common language or by the practices based on those implicit rules. The result will not

1.4 Imaginary Examples and Conceptual Analysis

look like the commonsense concept that I take as a starting point. I offer this list of properties and make no decisions now about which, if any, of them are essential for individuality. I approach the general question through a variety of concrete living entities, those we consider individuals and those that we do not. It may well turn out that there is no set of properties shared by every entity we call an individual.

Although I am reluctant to consider them as necessary and sufficient conditions, there are a group of properties that most of the organisms we consider to be paradigmatic individuals share. The relevant characteristics held by the higher plants and animals are:

i. spatial and temporal continuity,
ii. spatial and temporal boundedness,
iii. composed of heterogeneous causally related parts,
iv. development from a single cell to a multicellular body,
v. subject to impaired function if some of its parts are removed or damaged,
vi. ability to reproduce sexually, and
vii. genetic homogeneity.

These properties are, to a greater or lesser extent, common to a horse, a termite mound, an oak tree, a stand of bamboo, a slime mold plasmodium, a lichen, and a eukaryotic cell. Does this make them each individuals? Can we coherently extend our concept of individuality to all of them? These properties may come as a group, but that is not always the case. This is part of the problem in specifying exactly what is and is not an individual. Some entities have only some of these properties. Also, these properties can be held to different degrees even though the judgment of individuality is generally recognized as an all-or-nothing decision.

The commonsense notion of individuality is just that, a commonsense notion. We often think of a lichen as an individual thing. The lichen is composed of algae and fungus combined in symbiosis. This does not imply that biologists must accept the lichen as an individual or that commonsense folk ontology must accept other symbiotic unions as individuals. Even if it makes sense to begin with our folk ontology, our analysis of individuality may lead us far from where we began.

1.4 IMAGINARY EXAMPLES AND CONCEPTUAL ANALYSIS

Thought experiments and imaginary examples have a venerable history in philosophy.[6] Constructing thought experiments and puzzling over them is fun and difficult to resist. Unfortunately, they often mislead us and it is unclear what they contribute to philosophy.

> [P]eople think that they can conceive of a centaur, and in a very superficial way, they can, but too many questions remain unanswered, and no means exist for answering them. How many pairs of lungs does a centaur have, how many hearts? How are the circulatory and pulmonary systems of these creatures connected? What happens to food that has been digested in the human half of the centaur? Does it empty into the stomach of the horse half? Of course, we are not supposed to ask such questions of mythical creatures. (Hull 1989, p. 312)

I use very few thought experiments in this work. Instead, I rely on real examples. For the project I have in mind, thought experiments and imaginary examples more often confuse than clarify.

I am primarily interested in devising a system of individuation for living things capable of explaining unusual real cases as well as more common real ones. My goal is to construct a theory that can best accommodate the diversity of organisms and other living entities, not to derive the necessary and sufficient conditions for being a living individual in any possible world. Merely imagined possibilities are not particularly relevant to this project. I limit my examples to real living things for reasons that I explain below. This choice affects my arguments. It determines the kind of counterexamples that I need to consider, which will, in turn, affect the structure and scope of my arguments. Imaginary, though logically possible, examples do not provide direct evidence against a position on individuation that has been tailored to fit this world.

The abundance of real counterexamples that create tension in our commonsense conceptual boundaries renders imaginary examples inconsequential. The examples may not always be obvious, but they are plentiful, and unlike thought experiments, real examples come complete with the relevant background conditions either known or discoverable. Given the choice between a real example and an imaginary one, we should choose the real one.

The usefulness of a thought experiment depends on a correspondence between imagination and possibility, a relation subject to at least two common sorts of failure. We can apparently imagine an impossible state of affairs without recognizing its impossibility. And we may be unable to imagine something that turns out to be possible.

It is debatable what role thought experiments and imaginary examples should play in revising our concepts. We hope that our concepts help us to think about the world in a useful way. If a concept is inadequate to deal with an imaginary but possible example, that is a poor reason to reject it. Unless

1.4 Imaginary Examples and Conceptual Analysis

the concept is specifically intended to apply to every possible world, we have little need to hold our concepts to such standards. I more fully develop each of these three general reasons for favoring real examples and avoiding imaginary examples and thought experiments throughout the rest of this section.

Real organisms present puzzling examples that must be dealt with in a general treatment of identity. There is such a variety of living things that the interesting properties of the mythical creatures we invent may well be found in real organisms. After we construct a theory of biological individuality that can deal comfortably with the issues raised by actual organisms and living systems, perhaps we can discuss how to individuate imaginary creatures, but not until then.

My imagination is not as fertile as the living world. As J. B. S. Haldane said, "The world is not only queerer than we imagine, it is queerer than we can imagine." I could not have thought up the bizarre life cycle of the root-head or the strange reproductive strategies of the aphids.[7] I could not have imagined the myriad forms that living things assume or the Rube Goldberg life cycles they have cobbled together to reproduce and track their environments. This is not to say that once I heard about them I continued to be unable to believe that they were possible, but I am confident that I could not have thought them up. Because this is true, a theory of individuation based on my imagination and a layman's knowledge of the common organisms would not generate the more exotic problematic cases found in nature.

If something is actual, it is certainly possible. Thought experiments and imaginary examples almost never provide a clear description of background conditions, which are often crucial to determine what importance the example or experiment is to have. The background conditions for a thought experiment tend to be sketchy in a way that favors the point of view of the person who has formulated it or in a way that obscures some controversy. Because it is difficult to determine whether putatively possible thought experiments and imaginary examples are really possible or rest on an unrecognized contradiction, it is not as trivial as it initially sounds to say that a virtue of real examples is that we know that they are possible. Possibility is not as obvious a property of imagined but putatively possible examples. The connection between imagination and possibility is not as simple as it may appear.

We hope that we can imagine something if (and only if?) it is possible. Conceivability may be the best guide to modality we have, but why should human powers of imagination be a good indication of mind-independent possibility? We sometimes imagine states of affairs that we believe possible, but we later learn that they are impossible and sometimes we are unable to imagine real possibilities.

Our intuitions work best under normal circumstances. We are more likely to make mistakes like the ones above when we lack some important information, for example that 'Hesperus' and 'Phosphorus' are both names for the planet Venus. The further away from home we travel, the less reliable our intuitions are likely to be. We cannot and should not completely abandon our intuitions about possibility and impossibility, but we can try not to rely on them where they seem most likely to go wrong. If we get far enough away from normal, it is no longer clear that we can say with warrant that we have imagined a real possibility or dismissed a real impossibility and have not instead illegitimately relied on our ignorance of a contradiction we assumed possible.

Thought experiments and imaginary examples are particularly prevalent in the discussion of identity, especially personal identity, and have been for a long time.[8] A fair sampling includes the Ship of Theseus, Descartes's argument that his mind is distinct from his body, Marjorie Price's discussion of metamorphosing dogs, and a variety of commentators on the paradoxes of a person surviving an assortment of *Star Trek*-style transporter mishaps in which a person is broken into component parts and beamed elsewhere at the speed of light, or just her "blueprint" is, or one part goes one place and another goes somewhere else.

I do not know how many of these thought experiments describe real possibilities and neither do their inventors. Here are a pair of thought experiments that strike me as describing more or less impossible situations as possible. At minimum, neither convincingly establishes that it describes real possibilities.

> For this reason, from the fact that I know that I exist and that meanwhile I judge that nothing else clearly belongs to my nature or essence except that I am a thing that thinks, I rightly conclude that my essence consists in this alone: that I am only a thing that thinks. Although perhaps (or rather, as I shall soon say, to be sure) I have a body that is very closely joined to me, nevertheless, because on the other hand I have a clear and distinct idea of a body – insofar as it is merely an extended thing, and not a thing that thinks – it is therefore certain that I am truly distinct from my body, that I can exist without it. (Descartes 1984, Meditation Six)

Descartes reasons that because he is able to conceive of his mind without his body, his mind and body must be distinct. A more recent example of a more biological nature is provided by Marjorie Price.

> To determine the effects of the Martian atmosphere on higher animals, NASA sends Rover [a dog] to Mars. After a successful Mars landing and take-off, Rover returns to earth, where he is continuously observed for six months.

1.4 Imaginary Examples and Conceptual Analysis

> Film cameras record every moment of his existence. During this time, Rover undergoes a gradual change, so that by the end of the isolation period he is an amorphous mass of cells. Even the chromosomal constitution of his cells has changed: its nature is not identifiable as the sort to be found in members of any known kind of organism.... No one can deny that the entity in the isolation unit at the end of the interval in question, call it "Clover," is Rover, the object confined there six months earlier. (Price 1977, p. 203)

It is not obvious that Price's conclusion really follows from her thought experiment. More importantly, it is unclear whether she describes a process that could really happen to a dog.

Some thought experiments and imaginary examples are innocent enough. The innocent ones generally describe experiments that could be done but are not. Wilkes calls these merely imagined, "those experiments which are not *in fact* carried out in practice, but which could be" (Wilkes 1988, p. 3). It is not necessary to throw a particular beer bottle against a brick wall to be assured that it would shatter if we did. Pointing out that it could be done is enough. Somewhere between these unobjectionable thought experiments and Descartes's argument that he could exist without his body is the line beyond which we should treat thought experiments with suspicion.

Some thought experiments cross the line and become inadvertent literary fantasy. Literary fantasy makes a dubious basis for philosophical argument. Peter van Inwagen makes the case for omitting such tales from philosophical speculation.

> In my view, one may not use examples from fantasy in conceptual investigations. The reason is simple: the author of a fantasy has the power to confer "truth in the story" on known conceptual falsehoods. I could, for example write a fantasy in which there were two mountains that touched at their bases but did not surround a valley. *A fortiori*, the author of a fantasy has the power to confer truth in the story on a proposition such that it is a controversial philosophical question whether that proposition is a conceptual falsehood. (van Inwagen 1993, pp. 229–230)

Van Inwagen argues to disqualify literary fantasies from conceptual investigations. The movie *The Terminator* does not prove that travel to the past for the purposes of changing the future is a real possibility. It is a fantasy that includes this putative possibility and builds the story around it. To the best of my knowledge, *The Terminator* has not been cited as proof that time travel is possible, but imaginary examples with less detail have been used to support philosophical theories. These intentional or unintentional fantasies blur the distinction between truth and truth within the story. Van Inwagen intends to argue against intentional literary fantasy, but his argument should also hold for the inadvertent literary fantasies of the thought experimenter.

Thought experiments (or supposedly possible, though imaginary, examples) are dangerous because we construct literary fantasies without intending to do so. How much evidence can thought experiments provide for a philosophical theory when, like the theory they are used to defend, they may contain hidden inconsistencies? An imaginary example can be used to do this with unconscious ease, facilitated by the sketchiness and indeterminacy that characterize them. If it is possible to give a plausible perspective on what turns out to be an impossible landscape, it is also possible that a sketchy story contains an utter impossibility as an essential detail.

I end this section with a final illustration of the problems of the thought experiment. Imagine that you have somehow shrunk to the size of an ant and are being pursued by a giant ant (approximately the size you were before you shrank). It chases you to the edge of a lake, you strip off your clothes and swim to safety. If you are ignorant of the problems of scale this story involves, it may seem to be a real possibility; but it is not, given the physical laws that govern the world. The ant cannot chase you. It is writhing on its broken legs because at its current size its legs cannot support its weight. It is suffocating because the invaginations proportional to those that provided it with oxygen when it was its normal size are now inadequate because its mass has increased at a greater rate than its surface area. You are not faring much better. Molecular adhesion would hold your clothes so tightly against your body that you would not have the strength to remove them. Because of your new relation to surface tension you would have better luck running away on the surface of the water than attempting to swim in it.[9] Any world in which this supposed possibility occurred would have to have physical laws completely different from those that govern our world. Wilkes describes the problems underlying Parfit's imaginary example regarding personal identity:

> [I]n a world where we split like amoebae, everything else is going to be so unimaginably different that we do not know what concepts will remain 'fixed,' part of the background; we have not filled out the relevant details of this 'possible world,' except that we know it cannot be much like ours. But if we cannot know that, then we cannot assess, or derive conclusions from, the thought experiment. (Wilkes 1988, p. 12)

Most thought experiments leave the background conditions dangerously unspecified. Knowing that something is possible is very different from not knowing that something is impossible, even if it feels the same.

Possible, though imaginary, examples may be used as counterexamples if someone claims to have discovered the necessary and sufficient conditions for

1.4 Imaginary Examples and Conceptual Analysis

applying a given concept in every possible case. If, however, someone makes the more modest claim that a particular set of conditions are the necessary and sufficient conditions for applying that concept in the actual world, an imaginary example does not refute the claim. It does not undermine the claim that the concept functions well enough in the actual circumstances in which it is used.

Richard Gale makes this point well. He argues that thought experiments have been used as evidence that our concepts are inadequate because they are not applicable to every possible world, even those in which relevant details are radically different.

> What is perverse about these science-fiction thought-experiments is that they transport us, along with our present language-games and their forms of life, into the counter-factual world.... What they fail to realize is that in this world we would not want to play our old personal identity language-game, since there would be no point or value in doing so: the empirical presuppositions for doing so are not realized. (Gale 1991, p. 301)

Our concepts are developed in response to the way the world is. If things were quite different from the way that they are, people would probably not use the concepts we use. They would use ones that better fit the world as they found it. If genetic homogeneity had no relation to the individuality of an entity or to the traits it inherits, people would probably not have a concept for a form of individuality that depends on it. But so what? Why should our concepts be applicable to all worlds? As Gale points out, it is not fair to ask that they do apply.

Counterexamples can be real, imaginary but possible, or impossible and incoherent but presumably disguised. If what we are trying to describe is a situation completely different from the way things actually are, we may want to rethink the criteria we are using for our concepts, or rethink our conceptual scheme altogether. Perhaps a more refined or altogether different one would work better in a given (imaginary) circumstance. If the world were completely different, we would not necessarily speak, act, and think in the way that we actually do. Fair enough, but if you think of your words and concepts as tools, this should bother you as much as discovering that if you had to do a completely different job, the tool that is perfectly adequate for a task at hand is not appropriate for this other job that you have no need to do or interest in doing. This has no effect on the usefulness of the tool for the task at hand.

There are different ways to stretch a familiar concept. Some are more important than others. A real counterexample provides a better reason to modify a concept than an imaginary one if we are concerned that our concepts fit real cases we have not thought of yet. It is difficult to find necessary and sufficient conditions for the application of some ordinary language terms. Common sense may or may not pick out natural kinds.

For the most part, we do not freely choose our concepts or our conceptual scheme. What would it be like if we did? What we can do is rebuild and adjust this scheme as we go. The commonsense notion of what an individual is does not lead to clear answers when applied to living things that manage to fulfill only some of the criteria to some extent and some of them not at all or to a lesser extent than the paradigmatic multicellular individual. Our commonsense notion of individuality is not refined enough to function in specialized scientific situations. It runs roughshod over significant biological differences. Before returning to the subject of what concepts should replace our commonsense one, I explore the framework from within which this question should be answered.

1.5 WHAT IS IT?

In the chapters that follow I develop a new theory of individuation and persistence for biological entities. This makes this book a work of metaphysics as well as philosophy of biology. In this section I develop the metaphysical framework within which I will address biological individuality. I begin with the assumption that a living entity is a potentially finite three-dimensional persisting object. I do not argue for this assumption, but a skeptical reader should consider fully what is entailed by its denial. If this assumption is correct, a living entity is not a construction built up from momentary objects or time slices. Because it is potentially finite, a living entity can survive certain kinds of change, but there are other changes that it cannot survive. A living thing comes into existence and persists through time. For any living thing, there is some possible change that it can undergo but not survive.

This assumption is not foreign to common sense, nor is it philosophically innovative. It is an ontology with venerable roots in philosophy. Some philosophers and many biologists may bristle, however, at the temporal essentialism implicit in this assumption.[10] If an entity can endure only certain kinds of change, then the properties it has that cannot be changed are essential to it, meaning it cannot continue to exist if it no longer has those properties.

Essentialism is an unfairly maligned doctrine. A particularly noxious form of it based on the stereotypical morphology associated with a biological species has been taken to stand for all versions of essentialism. I will return to the subject later, but even before we have determined what the full complement of essential properties is for a given entity, we know that if a thing is not immortal there must be changes it cannot survive given its actual origin. If a magnolia tree is burnt to ashes does it continue to exist? Could a magnolia tree

1.5 What Is It?

become an animal? The answer to both questions is clearly no. If we accept those answers then we accept essentialism. It is essential to a magnolia tree that it continue to be a plant and that it not burn to ashes. This is essentialism of a sort; the only real alternative is an unattractive version of nominalism.

I can understand why a scientist or a science-minded philosopher might worry about essentialism. There is the reasonable concern that someone arguing for essentialism will endorse the idea of a kind of species essence, that is, a set of morphological properties definitive of individual members of a species in virtue of which that organism is a member of the class of similar organisms grouped together as a species. Further, this sort of species essentialism may be accompanied by belief that this sort of species essence is explanatory, or that it is a mysterious property that plays an important role in biological organization. A distaste for this form of essentialism has motivated philosophers as diverse in their views as David Hull (1965), Ernst Mayr (1975), and John Dupré (1993) to formulate arguments against it. I will not be arguing for biological essences of this kind. I will argue for a kind of essence that I think will be considerably more palatable.

Biology made a great leap forward when the metaphysical implications of evolution by natural selection were understood. Platonic archetypes in the mind of God or Aristotelian substantial forms were no longer used to explain the observed order of the biological world. Biology was recognized as a historical science. One of the more important changes that Darwin's theory brought to biology was that it made reference to essences illegitimate in a scientific explanation. I do not think that philosophers can understand why many biologists are opposed to essentialism of this form unless they realize what a fight it has been to purge these impediments to understanding the living world. But there is more than one essentialist doctrine in play and those who reject the forms of essentialism I mentioned above have little to fear from my position.

'Sortal' is a term coined by John Locke.[11] A sortal identifies a kind of thing. In contemporary philosophy the term picks out two distinct types of kinds. David Wiggins uses the term 'phase-sortal' to designate any sortal kind that is not a substantial sortal (Wiggins 1980, p. 27). 'Sortal' is used by some philosophers to refer to any kind that distinguishes some things from other things. Other philosophers use this term to refer not to just any sorting terms, but to only an important subclass of sorting terms, those that identify members of a *substantial kind*. A substantial kind provides the most basic answer to the question "What is it?" To distinguish the latter subclass of sortals from the other sortals, I will follow tradition by calling them *substantial sortals*. A sortal is a substantial sortal just in case a thing correctly identified under the sortal cannot cease to fulfill the criterion of identity associated with that sortal without ceasing to exist.

This distinction marks the difference between the sortal *well-dressed man* and the substantial sortal *human being*. A change of clothes does not annihilate a well-dressed man. He may no longer be well dressed, but he continues to exist (if you call that living). If *human being* is a substantial sortal, the change from human being to corpse is a different kind of change. It is a change that the subject of the change cannot survive.

If we examine any living thing, we discover that it is not just a "thing" or a "that"; it is a thing of some particular kind or other. It may be an oak tree or a sea urchin but it must be a thing of some kind. If this is true, each living thing is a thing of some particular kind, rather than just a thing *simpliciter*. For any particular thing, a substantial sortal identifies what kind of thing it is. To say that each thing is a thing of some kind does not yet rule out the possibility that the thing is a *bare particular*, something that can survive any kind of change while remaining numerically the same thing. It would be possible for a bare particular to be a member of some kind or other throughout its existence without being a member of the same substantial kind throughout its existence.

If there is a variety of change that a particular thing cannot survive, then that thing is not a bare particular. I do not think that there are bare particulars, but the reasons why there are none are instructive. The notion of a bare particular may not even be a coherent one, but to narrow my focus here, I will consider whether it is possible for a living thing to be a bare particular. The fact that each thing is a thing of a particular kind does not decide whether living things are bare particulars, substantial particulars, or whether there are some of both among living things. The statement that

1. *Every thing is a thing of some kind.*

is ambiguous between the two determinate positions below.

1a. *A thing must belong to some kind or other at every time it exists.*
1b. *A thing must be a thing of some kind or other and there must be some kind such that it is of that kind throughout its existence.*[12]

(1b) implies (1a), but (1a) does not imply (1b), so the two statements are not equivalent. (1a) is consistent with the existence of bare particulars. (1b) is not. A protean entity that meets the requirements of (1a), and any entity must, might migrate from kind to kind such that at each time throughout its existence it is an individual of some kind, but at one time it may be an individual of one kind and at another time an individual of another kind incompatible with the first. It is not necessarily the case that there is a sortal that is true of it at every time that it exists.[13]

1.5 What Is It?

A particular that fulfills (1b) is a substantial particular. Each living thing is a substantial particular because it is a living thing of a particular substantial kind, and cannot survive the loss of that kind. It is a fundamental feature of our commonsense ontology of living particulars that they can survive only some kinds of change. Other kinds of change result in the entity's death or annihilation. If my assumption that living things are potentially finite three-dimensional entities is correct, then there is at least one substantial kind that characterizes that entity throughout its existence.[14] Though this may seem to be an extraordinary jump from the assumption, it is not. If the living entity in question is even potentially mortal, there are some changes that would destroy it. The fact that the entity is mortal or potentially mortal means that there are changes that entity cannot endure. If there are changes that it cannot endure then that entity has essential properties without which it would cease to exist.

This assertion does not entail that we currently know what kinds of changes would or would not annihilate the entity, only that there are such changes. There may be monstrous living things of which we have never thought. How can we know what kind of thing such a being is? We may never know what substantial kind a given particular belongs to. We do know that it belongs to a substantial kind. I explain below how this follows from knowing that there are some kinds of change that the entity could not endure. Nothing I have argued for above entails that we are able to pick out the substantial kinds effortlessly or even with effort when confronted with any specimen of that kind. It is an empirical process and a fallible one by which we determine the substantial kinds for living things.

If there are changes that an entity could not survive, that entity cannot be a bare particular because a bare particular can survive any change. Indeed, each living thing is a thing of at least one substantial kind for the duration of its existence. When it ceases to fulfill the criterion of identity for an entity of that kind, it ceases to exist. This leaves an important issue unresolved. What are the substantial kinds?

An ontology of substantial individuals has some interesting implications. One of these is the existence of real, mind-independent kinds. E. J. Lowe, a supporter of an ontology of substantial particulars, writes:

> Individuals are necessarily individuals *of a kind*, and kinds are necessarily kinds *of individuals*. In consequence, I maintain that realism with regard to particulars or individuals – the belief, in my opinion correct, that they may exist independently of the human or indeed any other mind – implies realism with regard to sorts or kinds. (Lowe 1989, p. 5)

I think that Lowe is right, at least about natural kinds. If there are real individuals of a given substantial kind, then that kind is a real kind of individual.

This observation alone does not entail that there are real kinds, only that *if* there are real individuals then there are real kinds. As part of my argument that living entities are real, I will argue that there are real natural kinds. What living entities there are, an issue not to be confused with which living things we notice or consider important, is in no special way related to any sentient being's conceptual scheme or form of life. I accept Lowe's conditional about the relation between living things and substantial kinds. I am less sure, as he himself seems to be, about kinds of artifacts or other sorts of nonliving natural objects. Looking ahead to the development of my answer to the first issue raised by the existence of substantial kinds, I argue that there are real substantial individuals and real natural kinds corresponding to the substantial kinds of individuals. I will spend the next two chapters answering the second question as to which kinds are the substantial kinds for living things. Before that though, I will examine some abstract issues about an ontology of substantial kinds and substantial individuals.

Philosophy, unaided by empirical science, cannot determine which kinds are substantial kinds. But even before we know which sortals are substantial sortals, we can still determine what must be true of a substantial sortal. By definition, if F is a substantial kind, any thing that is an individual of kind F ceases to exist when it ceases to be an F.

A case of transubstantiation would be a case in which an entity, x, is of kind F (believed to be a substantial kind) at one time, and is later of kind G, also believed to be a substantial kind (where $F \neq G$), while remaining numerically identical. We have several options for dealing with such an alleged case that do not contradict the thesis that transubstantiation by living things is impossible. The first step is to determine whether the individual, x, survives the change of kinds or not.

1. *x does not survive the change from kind F to kind G.*

If x does not survive the change, there is no transubstantiation. There are many ways of not surviving the change. The individual may be annihilated, decay, or split into two or more distinct entities, each of which is sufficient for x not to survive the change. If x survives the change it is clear that something is wrong with our initial formulation of the case.

2. *x survives the change.*

There are several ways this fact can be reconciled with the general prohibition on transubstantial migration. It may be that

2a. *F is not a substance sortal after all.*

1.5 What Is It?

The evidence that x has survived may cause us to conclude that because x has apparently ceased to be an F without ceasing to exist, F is not really a substantial kind. Instead, F may be a phase-sortal. If x survives the change then perhaps F is one of the two varieties of phase-sortal.

Another possibility is that

2b. *x is not of kind F.*

Perhaps our initial assumption that x was of kind F is wrong. If this were true, x's transformation would not be a case of transubstantiation. Perhaps individuals of the substantial kind of which x is a member mimic the attributes of an individual of kind F for a part of its life.

The last possibility is that

2c. *x is still of kind F.*

It may be that x has been altered to make it look like an individual of kind G, a different substantial kind, yet still be an F. Alternatively, G may turn out to be a phase-sortal compatible with x being both of kind F and of kind G.

Because a genuine case of transubstantiation would involve a contradiction, all cases of alleged transubstantiation can be resolved in one of these ways. Even before examining the details of a particular case we know that it can be resolved through one of the above solutions. It is an empirical question which is the right solution to any particular case, but the possibility of transubstantiation can be eliminated before we know what the substantial kinds actually are.

The schematic account of an ontology of substantial kinds and individuals I have given does not adjudicate between competing systems of substantial kinds. It only tells us what must be true of any system of substantial kinds that we accept. If transubstantiation is impossible, each living entity is an individual of some determinate kind for the duration of its existence. What those kinds are is still an open question. In the next chapter I explore why biologists should be concerned about the concept of individuality and look at the attempts that some philosophers have made to account for biological individuality.

2

The Biological and Philosophical Roots of Individuality

In this chapter I begin by briefly exploring some of the ways in which individuality is an important topic for biologists, even those without a theoretical bent. I end the first section with a pair of case studies demonstrating the reality of the problem I hope to solve. In the following sections I sketch some representative philosophical positions about the substantial kinds for living things. I end this chapter by developing a theory of natural kinds that will effectively underwrite the particular ontology of living things that I will develop in Chapter 3.

2.1 WHY BIOLOGISTS (SHOULD) CARE ABOUT INDIVIDUALITY

There is an interesting and often weird body of writing on individuality by biologists. T. H. Huxley (1852) and Julian Huxley (1912) both wrote about it. So did Haeckel (1879). At the turn of the century there was a spurt of odd monographs exploring individuality and development through cutting flatworms into bits and in other ways testing the literal meaning of an individual as something that could not survive division.[1]

Biologists have been making sporadic attempts to clarify conceptual and empirical issues about the identity and individuation of biological entities for a long time. Conceptual concern about the necessary and sufficient conditions for fulfilling the concept of an organism seem to have become fashionable several times in the past and then faded again into obscurity without resolution. The concept of individuality has been at the center of concerns about how to distinguish a tightly integrated colony from an organism, issues about what kinds of entities are involved in natural selection, as well as the idea that if population biologists are to be able to do their jobs, they have to be able to decide how many things of a specific kind they are looking at. Sexually reproducing physically discrete entities are easy to count. Others are more

2.1 Why Biologists Care about Individuality

difficult. It is important that a suitable method of counting them be developed. For plants, colonial animals, clones, and symbionts, among others, the answers to these questions are far from obvious. Issues like these even have an impact on biologists averse to excessive theorizing. The fact that there is not a clear way to make a distinction between an organism and a tightly integrated colony affects those biologists working on other aspects of sponge biology, even those not directly tied to issues of individuality.

The primary model of selection is based on the organism (itself an unclear notion) as a unit of selection. The primary unit on which selective pressures are applied is the organism's phenotype. How can this model of selection and individuality be altered to account for natural selection at levels above and below the organism's level of organization? How can this model be adapted to the question of the "evolutionary individual," particularly among colonial or asexual organisms? I think that some of the questions about the units or levels of selection can be answered by a more detailed account of organism-level individuality. Issues about selective pressures on units above and below the level of the organism have also added to the confusion. Leo Buss (1987) attempts to explain organism-level individuality and developmental processes as artifacts of selective pressures on cell lineages. D. S. Wilson (e.g., 1997) and others have argued that insect colonies and other social groups can function like individuals.

As evidence that there are unresolved conceptual issues about individuality in biology, I would like to explore a couple of cases in more depth. Though there are many other cases to cite, I will limit my discussion in this chapter to two examples of the problems surrounding the current conception of individuality in biology. The first is the dispute over the status of an enormous clonal mass of the fungus *Armillaria bulbosa* discovered on Michigan's Upper Peninsula. The second is a debate about the distinction between the growth and reproduction of living entities.

In the April 2, 1992, issue of *Nature*, three biologists, Myron Smith, Johann Bruhn, and James Anderson, announced the results of research that they had conducted on Michigan's Upper Peninsula. They took samples of fungus from a number of sites in a hardwood forest and performed DNA tests on them. Based on these tests, which showed that samples taken from a large area were genetically identical, they mapped out the clonal mass.

> A region of 15 hectares yielded *A. bulbosa* isolates with identical mating-type alleles and mitochondrial DNA restriction fragment patterns, both of which are highly polymorphic within the species. We provisionally refer to this group of collections as 'clone 1'. (Smith, Bruhn, and Anderson 1992, p. 492)

Smith and the others think that clone 1 grew to its current size through vegetative growth. The majority of its biomass is below ground in networks of cordlike aggregations of hyphae called rhizomorphs. It probably weighs more than a hundred tons. Based on its size and estimates of its rate of growth, they think that this clone was established at least fifteen hundred years ago. Their article ends with the following conclusion.

> This is the first report estimating the minimum size, mass and age of an unambiguously defined fungal individual. Although the number of observations for giant plants and animals is much greater, members of the fungal kingdom should now be recognized as among the oldest and largest organisms on earth. (Smith, Bruhn, and Anderson 1992, p. 431)

They conclude that the fungal mass is an individual thing and also an organism. In fact, they seem to equate being an individual with being an organism, though this is a contentious claim not supported by the evidence they offer in its defense.

In the same issue of *Nature* that announced the discovery of clone 1, Clive Brasier wrote a response to their article.

> The suggestion of Smith *et al.* that clone 1 deserves recognition as one of the largest of living organisms, rivaling the blue whale or the giant redwood, invites closer scrutiny. The blue whale and redwood exhibit relatively determinate growth within a defined boundary, whereas fungal mycelia do not. . . . although clone 1's reputation as a champion genotype may yet be secure, its status as a champion organism depends on one's interpretation of the rules. (Brasier 1992, p. 383)

Brasier's main complaint with Smith's research is not with the claim that clone 1 is big or old, but with the claim that it is one thing. He points out that because it is a common occurrence with other species of *Armillaria*, clone 1 has probably become fragmented into several independently functioning components. This possibility allows Brasier to challenge the assertion that the available evidence proves that clone 1 is a gigantic organism. He bases his challenge on two points. It is possible clone 1 is not a continuous structure. Nor does it look like, or grow like, a higher plant or animal.

His challenge was answered by Johann Bruhn, one of the initial researchers:

> We're not talking about something like a blue whale that has a skin. My concept of an individual is a continuous, genetically uniform living structure. There might be gaps, but there is a strong possibility of continuity. (Svitil 1993, p. 70)

2.1 Why Biologists Care about Individuality

Two issues divide these camps. The first is an empirical question that remains unanswered. Is clone 1 physically continuous or has it become fragmented? The other dispute is conceptual, though its answer depends in part on the answer to the empirical question. Is clone 1 an organism? Is being an individual sufficient for being an organism?

Even with all of the empirical facts known, the status of this fungal mass as an individual is unclear. The debate between these scientists points to a more general problem that both Brasier and Bruhn acknowledge: models of organic individuality developed to describe the metazoan animals cannot be unambiguously applied to all kinds of living things because many of them are quite different in modes of development, growth, and reproduction. This is not an isolated problem.

Another dispute concerned with ambiguity in the definition of individuality is the distinction between growth and reproduction. T. H. Huxley defined the individual in biology as the total product from a sexually produced single-celled propagule. "The individual animal is the sum of the phenomena presented by a single life: in other words, it is, all those forms which proceed from a single egg, taken together" (T. H. Huxley 1852).

Daniel Janzen also believes that the folk concept of an individual has been confused with the individual of interest to evolutionary biology. For example, he thinks that selection does not act primarily on the dandelion plant that the layman thinks of as an individual. Janzen calls the unit that functions as an individual with respect to selective forces the evolutionary individual (EI). Because most dandelion seeds are asexually produced, Janzen conceives of a sexually produced dandelion and its genetically identical asexual descendants as a single scattered entity composed of numerous layman's individuals.

> At any time, it is composed of parts that are moving around ("seeds" that are produced by apomixis), growing (juvenile plants), dividing into new parts (flowering plants), and dying (all ages and morphs). Natural selection could just as well have produced an organism with all these parts in physiological contact, but in view of the type of resource on which the EI dandelion specializes, this alternative arrangement of parts is clearly optimal. (Janzen 1977, p. 586)

An evolutionary dandelion is similar to a tree with no branches or trunk. The parts of an EI do not have to be physically united or causally integrated, only genetically identical. This genetic identity allows the EI dandelion to respond as a unit to the forces of natural selection. Janzen's basic message is that the layman's dandelion is not the unit on which natural selection acts. Instead, the entire genetic individual, however scattered it may be, is the object of selection.

To summarize, Huxley and Janzen define reproduction as sexual reproduction. Growth is the increase in the size or number of cells with that genotype, however arranged. A single layman's individual's growth counts as growth, but so does the asexual reproduction and development of layman's individuals. They define an individual as the sexual propagule and all of its asexually produced daughter cells, however they are arranged. The motivation for this view is that asexual reproduction is analogous to the growth of a metazoan body, so the EI's total reproductive output should be counted as the output of a single entity. It is considered analogous because both a metazoan body and the asexual clones, however scattered, are investments in the production of propagules from a common genotype.

This model is not the only way to look at the distinction between growth and reproduction. John Harper, an English botanist, has developed a notion of organic individuality that includes two levels of population structure. Harper distinguishes between the genet and the ramet as distinct units in plant biology. The genet is the total reproductive output of a sexual or asexual propagule, however scattered. "An individual genet may be a tiny seedling or it may be a clone extending in fragments over a kilometer" (Harper 1977, p. 26). The ramet is the unit of clonal growth, the module that often continues an independent existence if severed from the parent plant. Strawberries grow by means of runners. A sexually produced plant puts out runners that develop into new plants. Each plant connected by the runners is a ramet. All of the plants taken together, even if they are no longer connected by the runners, compose a single genet.

Harper's definition of 'growth' is more restrictive than Janzen's because he does not consider asexual reproduction via a single-celled propagule to be a form of clonal growth.

> The distinction made here between reproduction and growth is that reproduction involves the formation of a new individual from a single cell: this is usually (though not always, e.g., apomicts) a zygote. In this process a new individual is "reproduced" by the information that is coded in that cell. Growth, in contrast, results from the development of organized meristems. Clones are formed by growth – not reproduction. (Harper 1977, p. 27n)

Janzen and Harper disagree about whether or not the single-celled "bottleneck" of asexual reproduction marks the beginning of a new individual.

Richard Dawkins thinks that Harper's distinction between growth and reproduction captures a significant fact about evolutionary innovation in developmental patterns missed by Janzen (Dawkins 1982, pp. 253–262). The difference is that the single-celled stage in either sexual or asexual reproduction provides a new developmental cycle between each generation whether those generations are sexual or not. This single-celled "bottleneck" allows for major innovations in development if that cell contains a mutation. He thinks that in the growth of the ramet, though mutations may arise, they cannot result in radical reorganization of the body plan. In light of this example, Dawkins thinks that the individual is best defined as the unit initiated by a new act of reproduction via a single-celled developmental "bottleneck."

In summary, for Harper and Dawkins, reproduction is the beginning of a new developmental sequence through the sexual or asexual production of a single cell. Growth is the development of organized meristems, whether physically continuous or scattered. An individual is the unit initiated by a new act of reproduction via a single-celled developmental "bottleneck." The motivation for this view is that the single-celled "bottleneck" allows for a mutation in that cell to make a radical change in the development of the resulting organism, whether that cell was produced sexually or asexually. One thing to note is that this view may treat distinct life phases as distinct individuals, so this will not always preserve our commonsense notions either.

Both of these problem cases involve atypical examples. The commonsense notion of biological individuality was developed with metazoan animals in mind. Unfortunately, it cannot be unambiguously applied to all kinds of living things because many of them exhibit strikingly different modes of development, growth, and reproduction. In the next section I look at what other philosophers have made of individuality and examine whether their views are adequate to deal with these kinds of cases.

2.2 PHILOSOPHERS ON LIVING ENTITIES

Philosophers from Aristotle forward have tried to explain the individuation and persistence of living things, but almost all philosophical accounts of the individuation and identity through time of living entities are based on out-of-date biology and deal with only the most commonplace of examples or with thought experiments that amount to little more than imaginary examples. A close look at real and unconventional examples would lead to substantial change in philosophical views. Philosophers such as Aristotle, Hobbes, Locke, van Inwagen, and Wiggins have tried to answer questions about the individuation and persistence of living entities, but for the most part these discussions have drifted away from current biological fact or never had much to do with real biology in the first place. In this sense, this project is an extension of an already accepted philosophical topic, but approaching it from within a more naturalistic framework with more attention to the relevant biology.

Before explaining my own view in the next chapter, I will examine the views of Aristotle, Locke, and Wiggins, both their particular suggestions and the philosophical motivations for supporting them. Because Aristotle was the first to articulate a complete theory of substantial kinds, I examine his view first.

Aristotle's metaphysical position in the *Categories* through the *Metaphysics* can be read in part as an attempt to find an explanation for the biological phenomena he observed that was consistent with Empedoclean science, which seemed to be unable to explain the complexity of structure or the repeatability of form in biological entities (Furth 1988). Despite differences in the justification he gives, in both the *Categories* and the *Metaphysics* Aristotle claims that individual plants and animals are the clearest cases of substantial particulars.[2]

In the *Categories* Aristotle says that substantial particulars are those things that are neither in a subject nor said of a subject. His examples of primary substances are individual men and horses.

> A substance – that which is called a substance most strictly, primarily, and most of all – is that which is neither said of a subject nor in a subject, e.g., the individual man or the individual horse. (*Categories* 5 2ª 11–13)

A secondary substance, e.g., *horse* or *man*, is the kind of a primary substance. A substance can survive through changes in some of its properties, or receive contraries. "It is, therefore, distinctive of substances that what is numerically one and the same is able to receive contraries" (*Categories* 5 4ª 10–11). Thus, the same man can be cold at one time and warm at another while remaining numerically the same substance. In the *Categories*, primary substances are treated as primitives, not open to further analysis.

In the *Metaphysics*, Aristotle claims that a substantial living individual is a hylomorphic composite of matter and form. The individual man or horse is still the ultimate subject even though the ultimate substance is now matter.

> The occurrence in the megascopic world of these endlessly repeated, specifically identical, highly organized, sharply demarcated, integral structures or systems (*sustema*, he calls them or *sustaseis*) – the biological objects which are the substantial individuals, each one a unitary individual entity or a "this," each one exemplifying over its temporal span a sharply defined complete specific nature or substantial kind – stands out as a remarkable fact of nature which invites explanation. "Invites," not "defies" – how *do* such entities come to take shape, out of the Empedoclean swirl of mixing and unmixing, clumping and unclumping? (Furth 1988, p. 70)

2.2 Philosophers on Living Entities

Empedocles' biological theory of mixing and unmixing cannot explain these phenomena. Mere mixing is inadequate to explain the complexity of organization found in living things. The substantial form is Aristotle's explanation for the transformation of mixtures of different kinds of matter into a complex unified biological object. Aristotle posits the form to explain this structure and unity. When a thing comes into existence, the form and matter together compose a composite being. When that being ceases to have that form, it ceases to exist. For Aristotle, the substantial form is causally responsible for the persistence of the living entity. The essence or substantial form played an active causal role in maintaining the integration of the matter that composed an organism. Within the weaker system of essentialism that I endorse, the essence does not play any such active role. It is merely a way of marking the beginning and end of existence, which is determined by biology, not occult essences.

Aristotle believes that each living thing is permanently endowed with a specific nature (which it shares with other things of the same kind) that leads to similarity in structure, function, growth, and reproduction. He identifies the *eidos*, or species, as the substantial kind.

The substantial form is the cause of the metabolic self-sustenance that constitutes the continuity of that being. The mechanisms of metabolic self-sustenance are found in all living things.

> For biological substances, the general operating principle for what counts as the persistence of "numerically one and the same" substantial individual through change across time, is, that which (as best we can scientifically determine) constitutes the continuity of metabolic self-sustenance, the *trophe di' hautou*, of an organic body. This is because the fundamental function, *ergon*, of a biological substance – the function presupposed by any other functions it might have, the "most prior" element of its definition – is to live, when and only when it ceases to *live*, then and only then it ceases to exist. (Furth 1988, pp. 156–157)

The substantial form is the basis for the living thing being the kind of living thing it is. It also constitutes the metabolic processes by which that thing maintains itself. This theory cannot help us with the kind of problem I explained in Chapter 1 about the difficulty of counting individuals or lives. If Aristotle were right, each living thing would be individuated by being a particular composite of matter and form, but this does not give us practical guidance about how to make that sort of determination when confronted with ambiguous cases.

John Locke's *Essay Concerning Human Understanding* contains elements of both a nominalist and a mechanist account of the individuation of living things, both of which are conscious rejections of the substantial forms Aristotle postulated. In the first edition of the *Essay*, Locke put forward a nominalist account of individuation consistent with the corpuscularian tradition of Boyle and Hobbes, in which matter is the only substance. Whereas Aristotle considered the death of a living thing to mark the end of a substantial individual, the corpuscularians thought of it as nothing more than a change in the accidental properties of the matter composing the living thing. Except for the creation and annihilation of matter by God, they viewed all change as being only relative to a system of classification or manner of speaking. Because of this, the corpuscularians did not take the problem of individuating material substances as a serious philosophical problem. According to this view, death does not mark a singularly significant event in the history of a body because life is no more than one accidental property among others. When we speak of it as if it were important, this is only relative to our system of classification. A living thing persists for as long as it continues to fulfill that sortal or be of that kind. The position is a nominalist one because the sortal term is based in an arbitrary definition based on observation, or a nominal essence.

> What are the alterations that may, or may not, be made in a *Horse*, or *Lead*, without making either of them to be of another Species? In determining the Species of Things by our abstract *Ideas*, this is easy to resolve: but if anyone will regulate himself herein, by supposing real Essences, he will, I suppose, be at a loss: and he will never be able to know when any thing precisely ceases to be of the Species of a *Horse*, or *Lead*. (*Essay*, Book III.iii.13)

Locke says that the same particular can instantiate many distinct species by satisfying different nominal essences. So Locke seems to favor the position that there can be a single particular with distinct nominal essences, though he sometimes seems to shift toward the position Wiggins and others endorse, that there can be more than one thing in a place at a time. There is support in the text for both interpretations. The fact that he is concerned with nominal rather than real essences gives his view the flexibility to accommodate multiple nominal species for the same object.

Locke distinguishes between a thing's real essence and its nominal essence (*Essay*, Book III.iii.15). The internal, but unknown, constitution of a thing that supports its discoverable qualities is its real essence. Locke thinks that members of the same nominal species may belong to different objective species if their distinct real essences cause the observable features of a common nominal essence. For Locke, the individual can persist by retaining the same nominal essence even if it changes real species. Because Locke takes nominal essences to be arbitrary constructions, his position makes the identity through time of entities depend on arbitrary human constructions. This account is then counterintuitive, and does not fit with the importance Locke places on the property of a continuing life.

2.2 Philosophers on Living Entities

Also contrary to this interpretation is the fact that Locke does not seem inclined to offer unnatural gerrymandered individuals in accord with equally gerrymandered nominal essences as examples, which he could if it were all a matter of free human choice. He does discuss cases in which living things vary from the putative essence of that kind of creature, but he does so as a jab at Aristotle. "The frequent Productions of Monsters, in all Species of Animals, and of Changelings, and other strange Issues of humane birth, carry with them difficulties, not possible to consist with this Hypothesis" (*Essay*, Book III.iii.12–16). Locke is right that these malformed creatures point to problems with the notion of a specific essence that must be embodied by every member of that species, but that is not Aristotle's notion. Even Aristotle made allowances for deformation. The failure to embody the form adequately is an explicable phenomenon within Aristotelian biology.

In the second edition of the *Essay*, Locke adds a mechanistic account of the individuation of a living body that may be incompatible with the earlier nominalist version. In this later account, the unity and persistence of living bodies can be explained in mechanistic terms without reference to a substantial form. He develops a realist conception by treating a living thing as analogous to an inanimate mechanism such as a watch. The mechanistic criteria Locke establishes are sufficient to individuate a thing of a given kind at a time. How does a living thing persist through such changes in constitutive matter without a substantial form to unify it? Locke gives two mechanist answers to this question, though he does not seem to distinguish between them.

The first criterion is the continuation of *a life*, whatever form that life may take, which does not rule out the possibility of an entity surviving a change of substantial kind if the individual of the second kind continues to be a living entity. The second criterion is the continuation of a life *conformable to that sort of plant*.[3] For a plant to persist, it must be a plant and be spatiotemporally continuous with an earlier plant that has persisted through time. An animal has similar persistence conditions. Identity is preserved if the entity at different times is unified, just as the parts of the entity at any single time are unified, by a single continuing life. The continuing life is a matter of mechanical organization, not an Aristotelian substantial form in addition to the material organization.

Locke thinks that the similarities between an inanimate mechanism and an animate one have limits. Both mechanisms are organized and composed of parts. But

> ... in an Animal, the fitness of the Organization, and the Motion wherein Life consists, begin together, the Motion coming from within; but in Machines the

force, coming sensibly from without, is often away, when the Organ is in order and is well fitted to receive it. (*Essay*, Book II.xxvii.5)

A machine can stop and then start again. Locke thinks that a plant or animal cannot. It is not clear whether Locke thinks this is a contingent or necessary truth. I will return to this subject later.

A mechanist interpretation is a good first step toward a realist theory of individuation and identity through time. The causal connection between parts is a matter of fact, independent of human observation or interest. Similarly, it is not a matter of convention whether or not parts of a body work together in a mechanical fashion. The mechanical organization of the living thing may be its real essence, but, according to Locke, it is still individuated by its nominal essence.

> It is an artifact of sorts – one grounded in long and careful observation (natural history) and so too, probably, in underlying facts about the structure of things – but for all that, an artifact, due primarily to the abstract ideas we construct. (McCann 1987, p. 63)

To continue the life of a living thing the appropriate organization of parts is required. Even in cases of death in which the body is left relatively intact, as in, for example, carbon monoxide poisoning, cellular-level structure is irrevocably disordered. According to Locke, the nominal essence sets the limits of appropriate organization. It is continuation relative to a kind of entity determined by an idea or nominal essence.

There seems to be a consistent realist position based in mechanical organization that Locke could take, but he does not take it. Instead, he retains puzzling aspects of his nominalist position from the first edition. This is particularly odd because he seems inclined to speak of ordinary living things and does not use the nominalist element of his theory to justify odd nominal creatures.

Ayers provides the clearest explanation for Locke's peculiar set of views.

> It is as if Locke had merely taken over an Aristotelian respect for the logical importance of life, having rejected the ontology upon which the respect was founded. It is as if, having been forced by the standing doctrine of forms to focus attention on the significance of life and death, he brought to bear on the issue an extreme nominalist or conceptualist theory which makes identity (except identity of the substance itself) relative to a nominal essence; but entirely failed to realize that nothing in that theory justifies treating loss of life as a change that is in any way special or peculiar in itself. (Ayers 1991, p. 220)

Perhaps when Locke speaks as if he is supporting a nominalist position, he either is intending or should be intending to make this distinction between the

2.2 Philosophers on Living Entities

real essence that we know nothing about and the nominal essence we derive from observation. But this should not be interpreted in a way that makes all nominal essences, however arbitrary, equal.

In many ways, David Wiggins's *Sameness and Substance* (and his earlier *Identity and Spatio-Temporal Continuity*) marks a renewal of an Aristotelian approach to identity and individuation, though without the substantial form playing an active causal role. Wiggins is a realist about substantial individuals and substantial kinds. Because of this, he rejects Locke's nominalism with respect to sortal kinds. He thinks that we can invent a sortal concept, but must do so in response to the constraints of our scientific or protoscientific observations of the world. Some of the further constraints he adheres to are as follows.

> D(iv): f is a substance concept only if f determines (with or without the help of further empirical information about the class of fs) what can and cannot befall an x in the extension of f, and what changes x tolerates without there ceasing to exist such a thing as x; and only if f determines (with or without the help of further empirical evidence about the class of fs) the relative importance or unimportance to the survival of x of various classes of change befalling its compliants (e.g., how close they may bring x to actual extinction). (Wiggins 1980, pp. 68–69)

D(iv) has much in common with Aristotle's view and the metaphysical framework that I sketched in Chapter 1. It is empirical and is based on the entity's persistence conditions. Another principle that he defends is D(v), which has a Lockean ring to it.

> D(v): f is a substance concept only if f determines either a principle of *activity*, a principle of *functioning* or a principle of *operation* for members of its extension. (Ibid., p. 70)

Wiggins thinks that natural kind terms are the best candidates for substance sortals. Nominal essences are ill suited for the job because

> they leave unexplained, not only the way in which our conceptions of the compliants of a sortal predicate can evolve while still being conceptions *of* the very same natural kind, but also the non-arbitrariness of that evolution. (Ibid., p. 78)

On this view, the identity and persistence of individual specimens of a natural kind are discovered by elaborating what can and cannot happen to a member of that kind. What can and cannot happen to a member of a natural kind is something that we discover *a posteriori*. Our initial beliefs about the scope of that kind are open to further amplifications or amendment.

> Consider the nineteenth-century discovery that the elvers *Leptocephali* were in fact the young of the species Conger Eel, or the humbler but in some sense proto-scientific discovery that tadpoles become frogs. (Ibid., p. 88)

We are not obliged to retain our biological mistakes in our conceptual framework. Our concepts can grow and change in response to empirical evidence. According to Wiggins the expansion of the criteria of identity for members of that kind is best explained in terms of a preexisting natural kind whose essence we uncover. He is careful to point out that the substantial sortals that he recommends for living things are open to revision in the light of empirical discoveries.

> Nothing in this argument here is meant to depend upon a certain conceptual conservatism into which no philosophical inquiry into substance and identity should find itself forced, viz. the supposition that one can tell *a priori* for any given sortal, e.g., the sortal *tadpole* or *pupa*, whether or not it is a substance-sortal or merely a phased sortal. Room must be found for the empirical and surprising discovery that there is something which is first a tadpole and then a frog. (Ibid., p. 64n)

Wiggins still suggests substantial kinds of the same nature as those presented by Aristotle. His examples include *human being, frog, oak tree*. In the next chapter I use empirical evidence from the biological sciences to argue that categories like these are inadequate to deal with a variety of puzzling cases and must be replaced with a more complex vocabulary of natural kinds. Also, considerations about the ontological status of biological species make them inappropriate to play the role that Aristotle, Locke, and Wiggins have assigned them. Whatever they are, biological species are not substantial kinds.

Aristotle, Locke, and Wiggins suggest different philosophical underpinnings for the individuation of living things. Nonetheless, they are in remarkable agreement that the substantial kind for a living thing is something roughly like its biological species. They also agree that empirical evidence plays an important role in making judgments about identity for living things. Aristotle devotes approximately a third of his writing to biology. Locke endorses an empirical approach to the discovery of nominal essences. Wiggins explicitly advocates a scientific approach to determining the substantial kinds. Their views, or at least their views about the substantial kinds for living things, are generally accepted.

For too long, the metaphysical study of individuation has been conducted as if the answer to the substantial kind question was known and all that remained was to find a proper philosophical foundation for it. Important advances in

2.3 Natural Kinds and Substantial Kinds

biology have, for the most part, been ignored. I do not fault earlier philosophers, at least those who wrote before such discoveries were made, but I think that a careful examination of biological theory and practice will undermine theories of substantial kinds previously limited to biological species.

The next chapter is my attempt to rectify this situation by examining biological theories of individuality. Unfortunately, the concept of individuality has become a contentious issue in contemporary biology, and no clear answer waits to be read from a textbook. Experimental and theoretical advances have exposed paradoxes in the commonsense notions of organisms and species and even in those notions as used by biologists. The missing pieces, which entities are the substantial living individuals and which kinds are the substantial kinds, will be found in the resolution of these paradoxes.

2.3 NATURAL KINDS AND SUBSTANTIAL KINDS

In section 1.5, I developed the abstract constraints on an ontology of substantial kinds. Until we know what makes a kind a substantial kind, important questions will remain unanswered. How can we judge whether a kind is substantial? To say that a substantial sortal picks out a kind that an entity cannot cease to be a member of without ceasing to exist only restates the question. It does not answer it. Living things satisfy many sortals. Are the substantial kinds real kinds rather than mind-dependent constructions? A substantial sortal must pick out a real kind if an individual of that kind is a real individual because the substantial kind marks off the persistence conditions for members of that kind. If the substantial kind itself is a nominalist construction, then the individuals delineated by it will be defined by nominalist criteria. In this and the following section I explain why substantial kinds are natural kinds and then give an account of real, empirically revisable natural kinds based on the concept of a pattern.

David Wiggins suggests that the best candidates for substantial kinds are natural kinds. He grounds substantial sortals in natural kinds because this supports the idea that the substantial kinds are real. Natural kinds are discovered by empirical rather than *a priori* means. Scientists and laymen alike are at least occasionally fallible about natural kinds. When we discover our mistakes, we revise our conception of the natural kind in light of further evidence. Any plausible account of natural kinds must be consistent with this phenomenon. Wiggins provides a good example of this revision. He describes the discovery that what had previously been considered a species of small eels, *Leptocephali*, were actually young Conger eels, not a distinct species. This empirical discovery led to the revision of the persistence conditions for that species of eel.

Though he does not explain exactly what criteria define a natural kind or which of the natural kinds are substantial kinds, Wiggins links substantial kinds to natural kinds. There are fundamental problems with the account of natural kinds that he accepts, so I offer an alternative account and then apply it to the particular problem of discovering which kinds are the substantial kinds for living things.

Wiggins endorses Hilary Putnam's theory of natural kinds. Putnam, and Kripke before him, argue for a theory of natural kinds connected to the causal theory of reference, though it is unclear whether this account is intended to be merely consistent with the causal theory or somehow derived from it. The virtues of their account make it an attractive option for Wiggins. Their description of natural kinds makes it possible to revise our understanding of a kind in light of new evidence. The causal theory of reference makes the connection between the natural kind term and the kind that term picks out a necessary one. This necessity underwrites the modal attributes that characterize substantial kinds.

> For the name to stand for a natural kind, everything depends on whether there is some nomological grounding for what it is to be of the kind. If there is, and if the predicate is worthy to survive as a natural kind term, then the holding of the relevant principles is nothing less than constitutive of its exemplification by its instances. (Wiggins 1980, p. 80)

A member of a natural kind must have the distinctive characteristics of that kind. To have those properties is to be of that kind. Because an entity cannot, by definition, be a member of a natural kind unless it has the characteristics of the kind, it must have those characteristics. The kind is a natural kind if there is some necessary connection between the attributes of that kind.

Despite its attractive features, the Kripke/Putnam theory of natural kinds is incomplete. It provides us with a way to fix the connection between a natural kind term and its extension once we have identified an exemplar of that kind, but it does not explain how to identify a kind. In the next section, I examine the theory of natural kinds associated with the causal theory of reference and point out its shortcomings. I will then provide an alternative account of natural kinds based on the concept of a pattern. This account of natural kinds has the virtues Wiggins recognized in Kripke/Putnam and can resolve a number of issues about natural kinds unexplained by their view.

Kripke and Putnam have developed an influential theory of natural kinds that has held sway for too long. I begin by looking at the *a posteriori* account

2.3 Natural Kinds and Substantial Kinds

of natural kinds they support. Because his work on natural kinds predates Putnam's, I will consider Kripke's position first.

According to Kripke, a *designator* is a name or description. A *rigid designator* is a designator that has the same referent in all possible worlds. For example, the name 'Thomas Jefferson' is used to pick out a particular entity, Thomas Jefferson. We can use that name to identify Thomas Jefferson in any possible world in which he exists. This does not mean that there could not be someone or something other than Thomas Jefferson that is called 'Thomas Jefferson' in a possible world. It does mean that we can use the name 'Thomas Jefferson' to refer to Thomas Jefferson in that possible world, regardless of what he or other things may be called in that world.

Kripke thinks that, in most cases, names are rigid designators. A rigid designator gets and maintains its meaning in the actual world through a causal chain of reference leading from the initial baptism of the object by ostension or description, which establishes the connection between the object and the name. The name is passed between users of the language. If the chain of transmission is of the right sort, later users refer by the name to the originally baptized object.

He believes that the causal theory of names suggests several essentialist theses – for example, that the actual origins of biological objects are necessary origins, and if a material object has its origins from a particular hunk of matter, it could not have had its origins from any other hunk of matter. A table could not be made out of ice if it is in fact made of mahogany. Some of a physical object's microstructural properties are supposed to be essential.[4] These essentialist claims appear to be consistent with the causal theory of reference, but Kripke offers no further argument for them or reason to believe that they are logical consequences of the causal theory.

Kripke thinks that the causal theory provides the best account of natural kinds. First, pick an exemplar of a kind by a linguistic term through some form of initial baptism. For example, "I call this stuff nitrogen." We then discover the nature of the substance we have identified through empirical investigation.

> I believe that, in general, terms for natural kinds (e.g., animal, vegetable, and chemical kinds) get their reference fixed in this way; the substance is defined as the kind instantiated by (almost all of) a given sample. The 'almost all' qualification allows that some fool's gold may be present in the sample. If the original sample has a small number of deviant items, they will be rejected as not really gold. If, on the other hand, the supposition that there is one uniform substance or kind in the sample proves more radically in error, reactions can vary: sometimes we may declare that there are two kinds of gold, sometimes we may drop the term 'gold'. (Kripke 1972, pp. 135–136)

This caveat allows Kripke to explain our fallibility in identifying substances. The true extension of the term 'gold' or 'plant' is determined by the real essence of the substances that were picked out by our baptism. It does not matter if we are aware of those properties or not.

His view explains that we can change our beliefs about the extension of a natural kind term – for example, excluding iron pyrites from true gold while still referring to the same kind, gold. The referent of 'gold' is not defined by the nominal essence that we associate with the term. It is determined by what other objects share the essential properties of the initial sample baptized as 'gold.' Even if our initial idea of gold was no more specific than "any yellow metal," we still have been referring to gold. This is true even if 'yellow metal' and 'gold' do not have the same extension. We do not change the meaning of 'gold' when we exclude fool's gold from its extension. Instead, we *discover* that iron pyrites is not gold. The extension of a term can be different from the nominal essence by which we identify it. Though we discover the essence of gold through empirical methods, the essence we discover is necessarily true of the entire extension of the kind. The essence determines the extension. A natural kind is discovered *a posteriori*, but it necessarily has the essence it actually has.

Kripke does not explain two crucial parts of his theory of natural kinds. He does not spell out how a scientist discovers or decides if a substance has been isolated in the initial baptism. Given the possibility that an initial baptism can misfire if there is no uniform kind there to be named, his account should explain what a substance is and how we are to distinguish a substance from nonsubstances.

The second explanatory gap is related to the first. How does a scientist decide or discover which properties of the examined sample are essential properties of the kind to which it belongs and which are accidental? It is easy to imagine a scientist examining an arbitrary physical specimen and discovering a variety of its properties – its chemical composition, boiling point, or color. This is clearly an empirical exercise, but Kripke envisions the scientist going further, discovering which of those properties are essential to the kind to which the specimen belongs. "Whether science can discover empirically that certain properties are *necessary* of cows, or of tigers, is another question, and one I answer affirmatively" (Kripke 1972, p. 128).

This procedure involves more than determining the properties of the specimen, and it cannot be a purely empirical process. There is nothing about the property itself as examined in the specimen that can be observed to be necessary. The project of determining the essence of the natural kind embodied in the specimen itself depends on the judgment that the specimen, or some

2.3 *Natural Kinds and Substantial Kinds*

significant portion of it, is a uniform substance with a common essence. Once the essence of a kind is established, the grounds for deciding whether any arbitrary sample is of that kind or not are clear enough and empirical, but by what means do we discover whether there is a kind there or not? Kripke does not explain how to fill the gaps in his theory beyond a few suggestions regarding origin and microstructure. Without knowing what a natural kind is, a scientist cannot differentiate between "impurities" and the substance because to do this presumes that the scientist knows what she is looking for. It is not clear how Kripke would have us proceed but this is an issue that must be resolved. Before I consider Kripke's view further I will examine Putnam's treatment of natural kinds to see if it fares any better.

Putnam uses the causal theory of reference to develop a similar account of natural kinds.[5] He is trying to refute the view that meanings are mental entities.

> Most traditional philosophers thought of concepts as something *mental*. Thus the doctrine that the meaning of a term (the meaning 'in the sense of intension,' that is) is a concept carried the implication that meanings are mental entities. (Putnam 1975, p. 218)

This theory of meaning rests on the assumption that knowing the meaning of a term is being in a certain psychological state. Having the same intension entails having the same meaning.

He tries to prove that psychological states do not determine the meaning of a term by means of a thought experiment. Assume that there is a place called Twin Earth. On Twin Earth the liquid called 'water' is not H_2O. It is a different liquid with a complicated chemical formula that Putnam abbreviates as XYZ. XYZ is indistinguishable from water at normal temperatures and pressures. If a contemporary visitor from Earth was told about the chemical structure of XYZ he would say "On Twin Earth 'water' means XYZ." A resident of Twin Earth in the analogous situation on Earth would say that on Earth the word 'water' means H_2O.

Putnam then asks us to imagine the same scenario occurring at a time before chemistry was developed on either planet. Neither the person from Twin Earth nor the person from Earth would be able to tell the difference between H_2O and XYZ because of their superficial similarity to one another and would assume that both of them were using the word 'water' in the same way. Putnam claims that even if a person on Earth and her counterpart on Twin Earth were in identical psychological states, the extensions of 'water$_{TE}$' and 'water$_E$' would still be different from one another. One would encompass all XYZ and the other would encompass all H_2O. "Thus, the extension of the

The Roots of Individuality

term 'water' (and, in fact, its 'meaning' in the pre-analytic usage of that term) is *not* a function of the psychological state of the speaker by itself."[6]

How, if not through a nominal essence or intension, does a natural kind term get its meaning? If meanings are not entirely in the head where are they? Putnam uses the causal theory of natural kinds to answer this question. Point to a glass of water and ostensively define it as water. Assuming that the sample of liquid you point to is similar enough to most of the stuff we call 'water,' then we have an ostensive definition of water. Bearing the right sameness relation to the just-baptized sample of water is necessary and sufficient for something to be water.

Putnam thinks that it is a matter of empirical investigation whether the "same liquid" relation holds between the sample and any other putative specimen of water. The relation between a substance and the name of that substance is rigid and does not change meanings between worlds. Microstructure usually determines the relevant sameness relation that must hold between the exemplar picked out by the baptism and any other putative member of the same natural kind.

It is not always the case, though, that the sample has a common nature.

> Another misunderstanding that should be avoided is the following: to take the account we have developed as implying that the members of the extension of a natural kind word necessarily *have* a common hidden structure. It could have turned out that the bits of liquid we call 'water' had *no* important common physical characteristics *except* the superficial ones. In that case the necessary and sufficient conditions for being 'water' would have been possession of sufficiently many of the superficial characteristics. (Putnam 1975, pp. 240–241)

If there were no common chemical structure to the stuff we called water, if it were just a heterogeneous variety of microstructures with common observable characteristics, then those properties alone might determine the essence and therefore the extension of the kind water. But, because water is actually H_2O, it is necessarily H_2O.

Putnam's account of natural kinds has the same shortcoming as Kripke's account. For a kind like water, the real work is to discover what constitutes the "same liquid" relation. Aside from saying that microstructure is often, though not always, essential to a kind, Putnam does not provide a means to determine the relevant "same as" relation. What prevents someone from determining that there are two kinds of water, some of it H_2O and some of it XYZ, from the same empirical evidence?

The causal theory of reference is no help in some real situations similar to the Twin Earth example. Zemach points out that there are many different

2.3 Natural Kinds and Substantial Kinds

compounds in any normal sample of water. Some of these we may feel comfortable rejecting as impurities, though there is no principled way to decide between a variation of a kind and an impurity using the causal theory of reference.

> [H]eavy water is commonly regarded as a kind of water. The same holds for aggregates of T_2O, HDO, HTO, and DTO molecules (the number of varieties is eighty, since in each case the oxygen can be either O^{16}, or O^{17}, or O^{18}). All these, we say, are different kinds of water. (Zemach 1976, p. 120)

There are similarities between these compounds and there are differences. Are they all part of the extension of 'water' or should some be declared to be impurities? Neither the causal theory of reference nor the theory of natural kinds associated with it gives us any guidelines for making such a decision.

Each halogen (fluorine, chlorine, bromine, iodine, or astatine) has seven electrons in its outer orbital. Each forms a salt when combined with sodium. Each forms a strong acid when combined with hydrogen. Do the halogens as a group form a natural kind or is each of the halogens its own natural kind or both?[7] A sample of fluorine functions equally well to fix the reference of either the element fluorine or the halogen group. For which of the two it is an exemplar depends on the modal properties that we recognize when making the initial baptism. Fluorine is a fine sample of the kind fluorine and it is also a fine sample of halogen. What determines whether all the halogens or merely fluorine are a part of the natural kind picked out by the "same-substance relation" that holds between the sample and other gases? The problem of determining the reference of the natural kind term may plague any theory that permits ostensive definition, but given the potential for confusion, some mechanism should be added to their theory to make clear the natural kind to which one is referring. A better theory might not depend so heavily on ostensive definition to underwrite the modal properties of a natural kind.

The causal theory of reference does not provide an adequate account of natural kinds. It does not explain why a natural kind should be set at one level of specificity instead of a higher or lower level or if there can be more than one natural kind that the same exemplar identifies. This may be a problem common to many theories of natural kinds. Beyond this problem, neither Kripke nor Putnam explains how a scientist decides which of the many properties of a kind are the essential ones, nor how to resolve disputes between scientists about natural kinds.

What makes their view attractive despite these flaws? Some characteristics of their view make it attractive as a source of substantial sortals. The substantial kinds must be real if the individuals of that kind are to be real, and their theory is a theory of real mind-independent natural kinds. Their theory is consistent with the fact that our beliefs about a natural kind can be fallible and change in light of empirical evidence while still being beliefs about that kind. A theory of natural kinds that replaces their theory should retain its virtues. A good theory would also go beyond the causal theory to explain how scientists discover essential properties and how to adjudicate disputes between scientists about what these properties are. It must also be able to answer the referential challenge set by heavy water of various sorts and the halogens and explain what to do with cases in which there seem to be higher and lower essential natures. Do we need to decide between them, and if so, on what principles should that decision be made?

2.4 PATTERNS AND NATURAL KINDS

In this section, I present an account of natural kinds that preserves the virtues that made the causal theory attractive as a source of substantial kinds while avoiding its shortcomings. A natural kind is a pattern in nature.[8] A pattern is a candidate for pattern recognition (Dennett 1991, p. 32). A pattern's existence does not depend on its recognition as a pattern. There is no contradiction in an unobserved pattern, but it must be potentially recognizable. That a pattern must be potentially recognizable to be a pattern naturally raises the question of who or what does the pattern recognition. I know of no good reason to assume that human observational powers set the limits of real patterns. A creature with a different set of sense organs, cognitive capacities, or interests might perceive patterns that we do not or even cannot perceive. The pattern would be there even if it were unobservable by human beings.

Before discussing the relation between patterns and natural kinds, it would be good to get a clearer idea of what constitutes a pattern. Dennett clarifies the idea of a pattern with an idea that he adopts from a definition of mathematical randomness.

> A series (of dots or numbers or whatever) is random if and only if the information required to describe (transmit) the series accurately is incompressible: nothing shorter than the verbatim bit map will preserve the series. Then a series is not random – has a pattern – if and only if there is some more efficient way of describing it. (Dennett 1991, p. 32)

A phenomenon containing a pattern can be compressed by using that pattern. This idea is applicable to a far broader range of applications than computer data compression. Describing an entity as a specimen of gold or as a functional

2.4 Patterns and Natural Kinds

individual can take the place of an exhaustive explanation of an object in terms of more basic properties because the information can be compressed.

Human beings recognize a huge number of patterns in nature and use them to compress information when we form concepts or describe phenomena. The metamorphosis of a particular tadpole into a frog is unsurprising because it is in accord with a pattern recognized in past episodes of development. It is a regularity in nature. Where we find the one property we become accustomed to finding the other properties we have learned to correlate with the first one. This is a fallible process. We sometimes find that some of our assumptions are false or that properties that we thought must be coextensive are actually separable.

We treat some of the patterns we discover as natural kinds. What distinguishes the patterns identified as natural kinds from other patterns? A natural kind is a shorthand for a complete description of particular properties of a particular object or process. If we describe a hunk of matter as an organism, this description tells us nothing about many of the unique properties of that entity, but it does tell us a lot about how we could expect that entity to behave or change through time and other kind-specific properties it is likely to display.

Describing that same object as *an animal that would fit through a nine-inch hoop* is also to use a pattern or simplified generalization to describe it. We could make a number of valid predictions about it based on this property. But *animal that would fit through a nine-inch hoop* seems to be neither a natural kind nor a substantial kind. What makes a pattern in nature one that we are willing to accept as a natural kind or a substantial kind?

A variety of objective and subjective variables distinguish between the patterns we recognize as natural kinds and those that we do not. Some patterns, *animal that would fit through a nine-inch hoop* among them, allow too much interference to function as natural kinds. The creatures lumped together by this kind are quite heterogeneous. Such a pattern may be better than a bare guess but it may not be robust enough to justify paying much attention to it. A robust pattern is one that identifies a group of phenomena that share important causal or lawlike similarities. We may also discover another pattern that makes the same sorts of predictions and allows us to do so more accurately or with greater ease than the initial pattern. These criteria are a combination of subjective and objective standards.

We do not freely choose which patterns we recognize or which patterns we recognize as important. We do not freely choose our interests or perceptual abilities, and both of these influence which patterns we find important. If we are honest, it is obvious that we do not choose what kind of ontology to accept. Because the individuals we recognize are delineated by the kinds we recognize we do not freely choose which kinds we recognize either.

Our initial guess about the persistence conditions of a kind of entity are often wrong. So we revise them in response to empirical evidence or theoretical developments. This process continually takes place with many of our concepts and is not always orderly. Sessile animals may initially be identified as plants, or morphologically distinct life stages may be mistaken for distinct biological species. We may discover that a pattern of regularities that we thought would work for one sort of organism actually works better for another for which it was not designed. Organisms that we try to group together by their common properties may turn out to be heterogeneous enough to warrant placing them in two or more separate groups. There is no reason to suppose that this process will be orderly or that mistakes will not be made. Eventually, we find some patterns and associated identity and persistence criteria that are robust enough to survive close scrutiny. These are regularities like all mammals nurse their young.

The role of these patterns in scientific explanation depends on the nature of scientific explanation. If an explanation is no more than an abbreviated description of what happened or what will happen, then these patterns function as the abbreviations. Most philosophers of science believe that there is more to scientific explanation than making accurate predictions. It is contentious what that additional element is. The two most plausible views are that a scientific explanation must explain the causal mechanisms involved in a particular case or it must subsume the particular case under general causal laws or lawlike regularities.

> If the universe is, in fact, deterministic, then nature is governed by strict laws that constitute natural regularities. Law-statements describe those regularities. Such regularities endow the world with patterns that can be discovered by scientific investigation, and that can be exploited for purposes of scientific explanation. To explain an event – to relate the event-to-be-explained to some antecedent conditions by means of laws, is to fit the event into a discernible pattern. (Salmon 1984, p. 17)

Under this model of explanation, particular causal interactions are conceptualized as instantiations of general laws. I favor the other main alternative, which is to treat the general laws as the generalization of regularities actually exhibited by particular causal processes. The generality of the law is explained by the similar causal mechanisms in the cases that fall under it. The law is constituted by its instances. The most important patterns in biological phenomena will be those that reflect the predictability of similar causal mechanisms. Where there is a clear and stable pattern, there is probably an underlying causal structure.

The issues discussed above help explain which of the welter of available patterns we recognize as natural kinds. Two kinds of factors are involved.

2.4 Patterns and Natural Kinds

1. *Pragmatic considerations.* How easy is the pattern for us to recognize? Our ability to recognize a pattern will be tied to our perceptual abilities and limitations, our interests, and our access to the phenomenon.
2. *Explanatory power.* How much does the pattern allow us to explain or predict? The explanatory power of a pattern will tend to vary in direct relation to the generality and predictability of the causal regularity it represents as well as the relation between that causal regularity and other causal regularities.

The interaction of these factors determines which patterns we recognize as natural kinds. Neither of these factors, alone or in combination, represents anything like freely choosing an ontology of natural kinds.

The fact that there are subjective factors – for example, perceptual abilities or interests – involved in our selection of particular patterns in nature as natural kinds may appear to undermine the objectivity needed for these kinds to serve as substantial kinds for real individuals. If our choice of kinds is partially based on subjective criteria, are the individuals picked out by those kinds subjective too? Not necessarily. If we adopt a system of natural kinds that involves subjective criteria this does not entail that our choice results in subjective kinds if our subjective choice is between objective kinds.

Creatures with different interests or perceptual abilities could observe the same phenomena and recognize patterns and natural kinds that we cannot. Such beings could recognize a different ontology of individuals by using those natural kinds as substantial kinds. The abstract framework of individuals and substantial kinds that I described earlier in this chapter can be partially filled in different ways. Because any ontology of substantial kinds and substantial particulars must be based on natural kinds that are themselves based in objective patterns and causal structures, each of these ontologies represents a part of a complete ontology of real individuals.

My pattern-based theory of natural kinds retains the virtues of the Kripke/Putnam position. For Kripke and Putnam, natural kinds are discovered empirically, support modal claims, and are revisable. Issues unresolved by the causal theory include how we discover natural kinds and how scientists decide which of the archetype's properties are the properties essential to that kind. Within the view that I endorse, the patterns on which a natural kind is based are discovered empirically. A scientist does not examine an archetype and discern which of that sample's properties are essential to it. Instead, the scientist checks for relevant similarities between a sample of the putative substance and other samples and tries to determine which of that substance's properties causally underlie other of its properties. There may be no such property, in which case there is evidence that the sample is not homogeneous. If there are important similarities, then there is reason to investigate further. This is a flexible process that is not tied to the nature of any one particular exemplar. It also allows for us to account for how a scientist discovers essential properties of a substance through comparing different samples rather than direct observation. This method for discovering natural kinds is fallible and revisable. If a pattern does not ultimately prove to be as useful as was initially thought, or a better alternative is discovered, the properties associated with that natural kind can be adjusted in light of empirical evidence.

A pattern-based theory of natural kinds can also support the real essential properties necessary for a theory of substantial kinds. A natural kind represents a causally important pattern that strikes us as important and functions well in our scientific explanations. It picks out a group of individuals that fit the pattern. It is in virtue of fitting the pattern that individuals are identified as members of that natural kind.

Kind membership hinges on meeting the criteria that characterize that pattern. This fact alone does not make those properties essential to members of the kind. They are essential properties only if the identified kind marks the persistence conditions for individuals picked out by that kind. If the kind is a substantial kind then this condition will be met. It will have been met through the revision of the criteria for kind membership in light of empirical evidence until the persistence conditions for the kind are in equilibrium with the empirical evidence that the individuals of that kind continue to exist so long as they retain the properties of that kind and cease to exist when they lose any of those properties. Only then will modal claims about individuals that belong to that natural kind based on the properties of that kind be justified.

The pattern-based theory of natural kinds can support an ontology of substantial kinds and real individuals whose persistence conditions are established through those kinds. It retains the virtues of the view of natural kinds endorsed by Kripke and Putnam, while providing a more complete account of how natural kinds are discovered and revised. Because my theory of biological individuation involves several different but often overlapping substantial kinds of living entity in its ontology, the view I take on natural kinds must be compatible with a certain kind of ontological pluralism.[9]

A popular form of pluralism is the claim that the same individual may be conceptualized within different conceptual schemes. It is true that the same individual can be an individual of different kinds at the same time; but that individual can only belong to one substantial kind, because the substantial kind marks the limits of its existence. The only exception to this rule would be that an individual could belong to two or more substantial kinds if those kinds have exactly the same persistence conditions. If an individual could simultaneously be a member of different substantial kinds with different

2.4 Patterns and Natural Kinds

persistence conditions, then, if it continued to meet the persistence criteria of one substantial concept and not that of another, that entity would both exist and not exist. Because it is impossible for a thing to both exist and not exist at the same time, this option leads to absurdity.

We can approach the issue of pluralism in another way. The existence of one pattern in a phenomenon does not preclude the existence of another pattern in the same phenomenon, even if each of these patterns is the basis of a natural kind. The same parts can compose more than one object simultaneously. The pattern-based theory of natural kinds is compatible with an attractive form of pluralism. The discovery of one natural kind of living entity, for example, the *genetic individual,* does not preclude the existence of a member of the natural kind *functional individual* even if functional individuals sometimes overlap with genetic individuals. In Chapters 3 and 5 I describe these distinct kinds and the relation between them. I describe real cases in which the same group of cells simultaneously compose a functional individual and a genetic individual – for example, a non-twin metazoan animal composed of genetically homogeneous cells. The individuals picked out by the distinct substantial kinds may differ only in modal properties but they are still distinct individuals. For the reasons I discuss in the paragraph above and in the section about relative identity in the Appendix, even the complete overlap of matter does not entail identity because of the different persistence conditions associated with individuals of those kinds.

This form of pluralism is compatible with real substantial kinds and real individuals identified by those kinds. Patterns can overlap without impugning the reality of the overlapping patterns. A more specific pattern can be nested within a broader pattern without implying that any of the patterns are unreal. Natural kinds and therefore substantial kinds can be based on these types of patterns without conflict, provided that individuals identified under a substantial kind are not identified with individuals identified under a different substantial kind with different persistence conditions. Pluralism of this sort preserves the logical features of an ontology of real substantial individuals and real natural kinds at the expense of a sparse ontology, but I think that it is worth the price. Such a system of natural kinds allows us to describe more fully the many kinds of living entities that exist. A rich ontology of living things is not a sin if there are many kinds of living things.

3

Individuality and Equivocation

> The whole question seems to turn upon the meaning of the word 'individual.'
>
> T. H. Huxley, 1852

3.1 PARADIGM INDIVIDUALS: THE HIGHER ANIMALS

I assume that there are no real paradoxes in nature. An apparent paradox is a problem to be solved. To resolve the paradoxes of individuality I discussed in Chapters 1 and 2, I need a more refined vocabulary of individuation to describe and classify the diversity of life cycle and modes of growth and reproduction found in nature. In this section, I begin to develop this vocabulary by examining the properties of an adult higher animal that make it a paradigm of biological individuality.[1]

'Higher animal' refers to those animals that are a lot like human beings in that each is multicellular, composed of diverse types of cells, which in turn compose a variety of tissues and organs. 'Higher animal' is not a rigorous term, but its use will become clear when I detail the properties commonly associated with the higher animals. Most higher animals reproduce sexually and only sexually. Each higher animal develops from a single cell, which divides by mitosis into a group of cells that develop into its adult form.

Although a higher animal has the properties commonly considered relevant to individuality, and exhibits them to the greatest extent found in nature, I do not argue that a higher animal is a living individual of the highest degree or that it is a paradigm individual. I do not argue that the properties I list below are the necessary and sufficient conditions for biological individuality. Instead, I use this section to explore in detail the properties considered relevant to biological individuality in all its varieties. In the next section I explain why we should parse these various properties into distinct concepts of individuality that can be

3.1 Paradigm Individuals: The Higher Animals

separated from one another not only in theory but in living creatures. Because these concepts are all embodied in the typical higher animal, the observer frequently does not recognize distinctions between them or lumps them together as a single concept, even though few other living things have all of them.

To avoid talking about the typical higher animal in general, I will use Lucy, a ring-tailed lemur, as an exemplar for all the higher animals – though a brook trout, a pygmy rattlesnake, a narwhal, or a human being would work just as well. All of them are higher animals that share the properties discussed below.

Lucy is a *particular*, a material object.[2] As such, she is not a universal, or a class. Classes have members. Lucy does not have members. She is composed of parts, which are also particulars. A class can also have parts, but a class cannot have a particular of any sort as a part. For example, the class of all lemurs north of the equator has each particular lemur north of the equator as a member. This same class has the class of all lemurs north of the Tropic of Cancer and south of the Arctic Circle as a part. A particular can be a member of many different classes, but it is neither a class nor is it a part of one. Even a class with only one particular as a member is not identical to that particular. Lucy is the sole member of the class of lemurs mentioned by name in this book and one member among many of the class of lemurs north of the equator, but she is not either of these classes. A universal can have instances. No particular can have instances. Because Lucy is a particular this means that she cannot have instances and is not a universal.

Lucy is *spatially localized*. A spatially localized entity occupies some, but not all, space. At any given time there are places that she is and places that she is not. Although Lucy is spatially bounded we cannot infer that her spatial boundaries are precisely specifiable. She may have vague boundaries. A puff of smoke has spatial boundaries too, but we cannot pinpoint a precise nonarbitrary boundary between the puff and the surrounding air. It may be possible to specify the edges of the puff roughly to the inch, but beyond that point it quickly becomes an exercise in arbitrary boundary drawing. It is arbitrary because there is no principled reason to choose one such precise boundary over a range of other precise boundaries.

Lucy, however, does seem to have relatively sharp spatial boundaries when viewed with the naked eye. The apparent sharpness of her spatial boundaries is partially a result of our perceptual abilities. If we were able to view her subatomic structure she would appear less sharply bounded. Under those conditions, if current physics is approximately right, she would resemble a cloud or a puff of smoke and her spatial boundaries would be as vague. This vagueness does not mean that Lucy has no spatial boundaries, only that it is impossible to specify those boundaries beyond some point without being arbitrary.

Lucy is also temporally *localized*. There are times when she exists and times when she doesn't exist. She has a beginning in time and will have an end in time. But as with spatial boundaries, it may be impossible to specify precisely when a lemur begins to exist or stops existing without arbitrarily choosing one moment. This vagueness should not undermine our confidence that Lucy has temporal boundaries because there are clearly times when she did not exist and there will clearly be times when she no longer exists and these conditions suffice to ensure temporal boundaries.

Not only does the ring-tailed lemur have spatial and temporal boundaries, she is *spatially and temporally continuous*. Her spatial boundaries at any given time bound a continuous lemur-filled space. This space, however, does not include all of the space within the boundaries of her outline. For example, there is presumably a hollow tube of non-lemur-filled space that runs from Lucy's mouth through her digestive system. The contents of this tube are not a part of Lucy either. Lucy is not like the British Empire of the nineteenth century, which, though spatially bounded, was at any given time composed of a number of spatially discontinuous parts.

Quine does not think that Lucy really is spatially continuous.

> The territory of the United States including Alaska is discontinuous, but it is none the less a single concrete object; and so is a bedroom suite, or a scattered deck of cards. Indeed every physical object which is not subatomic is, according to physics, made up of spatially separated parts. (Quine 1950, p. 624)

Because she is composed of subatomic particles with gaps between them she is, in fact, a spatially discontinuous particular. When we speak of an ordinary particular being spatially continuous we generally mean that there are no macroscopic gaps between its parts, and ignore the microscopic gaps between its subatomic parts. I call this degree of spatial continuity *relative spatial continuity*. Lucy is relatively spatially continuous. I call the more restrictive criterion, that there are no gaps of any size between parts of any size, *absolute spatial continuity*. Because all living things are composed of gappy subatomic parts, no living things are absolutely continuous. If absolute spatial continuity is a necessary condition for being an individual, there are no living individuals, because there are no subatomic living individuals and anything composed of more than one subatomic part has at least microscopic spatial gaps between its parts. We may take this as proof that absolute spatial continuity is not a criterion of biological individuality, because applying it would entail that there are no living individuals and one of our assumptions is that there are such individuals. In either case, the importance of spatial continuity will, I think, be eclipsed by two varieties of causal rather than spatial relations between the parts of a living individual.

3.1 Paradigm Individuals: The Higher Animals

Lucy is also *temporally continuous*. From the time she begins to exist she remains continuously in existence until she ceases to exist. She does not pass in and out of existence like a river that goes dry and then flows again. Living things are not temporally gappy, though they may have vague temporal boundaries.

She has other properties that are just as important as having boundaries and being more or less continuous in space and through time. Something more makes Lucy an individual lemur. The properties I have described so far are not all of the relevant properties. If the properties discussed so far were sufficient conditions for being a living individual, Lucy would become a part of a new living individual each time she bumped into another lemur. If this were all that were involved in being a living individual, Hands Across America would have marked the formation of an enormous new living individual – which it did not. Two types of causal connection prevent such a formation.

In an attempt to answer a different, though related, question, "In virtue of what do parts compose something?," Peter van Inwagen considers and then rejects a series of relations – contact, fastening, and fusion among them. He does not believe that these relations are sufficient to join parts together into a new individual, because these spatial or superficial causal relations are insufficient to bind any two living things into a new living individual composed of both of them.

> Any bonding relation that can hold between any two moderate-sized specimens of dry goods can (I should think) hold between two human beings, and it is pretty clear that one cannot bring a composite object into existence by bonding two human beings – or two living organisms of any sort – to each other. (van Inwagen 1990, p. 62)

Although it is not clear at all that if two living things do not compose something via some spatial relation, *no two things*, living or not, can compose something by that relation, van Inwagen is right that a new composite individual is not brought into existence by gluing hamsters or snakes together.[3] What properties do the parts of the snake have that bind them into a single being that they do not share with the parts of a hamster even when the snake parts are in contact with or fastened to the hamster parts? It is the causal rather than spatial relations that hold between them. Causal processes of the relevant sort usually are transmitted through spatially contiguous parts.

Lucy, our exemplar of the living individual, is *composed of heterogeneous causally related parts*. Parts may be heterogeneous at one level of organization and homogeneous at another. Buildings can be very similar even if each is composed of a different sort of brick, rendering them heterogeneous at the brick level and homogeneous at the building level. The same sort of brick can also be used to build different kinds of structures that are homogeneous at the brick level and heterogeneous at the building level. Lucy's parts are heterogeneous at a variety of levels of organization. In her body, cells of different kinds compose her varied organs and tissues. The heterogeneity of her organs is more or less obvious. A heart is different from a kidney. Less obvious is the variety of cell types represented in her body. The numbers are only approximate, but she is composed of at least 120 different cell types and probably many more. The exact number is unknown (Bonner 1988, p. 133). This heterogeneity is to be contrasted with the homogeneity illustrated by a colony of green algae, which, however large it may become, is still homogeneous at the level of the cell because it is composed of cells of only one type.

Her heterogeneous parts are bound together by two different kinds of causal relation, *causal integration and causal cohesion*.[4] Because direct causal interactions are generally between spatially contiguous parts, it is easy to confuse the two kinds of causal unity. They are distinct. Mishler and Brandon define 'integration' as active interaction between the parts of an entity. "Does the presence or activity of one part of an entity matter to another part?" (Mishler and Brandon 1987, p. 400). The operation of Lucy's kidneys has an effect on her blood. The relation between her pituitary gland and hypothalamus affects her growth. Her skeletal muscles and bones have effects on one another. Secretions from some of her cells affect the activities of other cells, which in turn affect others. Her central nervous system oversees much of this integration. Not only is Lucy physically integrated, she functions as a *unit of behavior*. She acts as if she has a single control center with her component parts working together. Her legs and head do not seem to be pitted against one another. They function together.

Causally interacting entities display varying degrees of causal integration. The degree of coordination among diverse parts is what causal integration measures. Lucy's organs are causally integrated; so are the parts of a good pit crew or a state legislature. The difference is one of degree, not of kind. This does not make it a trivial difference, but it does make it difficult to specify exactly how complex the interactions between parts needs to be for them to be parts of a causally integrated entity. I will return to this question later in the chapter.

The fact that Lucy would typically suffer impaired function if enough of her parts were removed or damaged is related to the causal integration of those parts. Small loss of insignificant parts might not have an adverse effect on her ability to function. Richard Dawkins describes one of the attributes of an organism as "the quality of being sufficiently heterogeneous in form to be rendered non-functional if cut in half" (Dawkins 1982, p. 250). Being cut in half is one of the more extreme forms of having parts removed or

3.1 Paradigm Individuals: The Higher Animals

damaged, and any higher animal, Lucy included, would certainly be rendered nonfunctional (dead) by bisection. Bisection is only one of the many ways of removing or damaging parts from a higher animal, and death is only the most extreme form of impaired function. A multitude of other less-than-bisecting cuts would also remove enough parts of a higher animal to render it nonfunctional and an even larger number of such cuts would impair its function. In this respect, Lucy is different from many (relatively) spatially continuous but modularly constructed living entities like Siamese twins connected at the tips of their thumbs, who could be split between the shared thumb tip and be little the worse for wear. As with many of the properties I have been discussing in this section, the impairment of function has different degrees. Do strawberry plants connected by an above-ground runner suffer impaired function when the runner between them is severed? Not in any apparent way. Does a zygote suffer impaired function if it is separated into two cells that develop into monozygotic twins? The degree of impaired function a living entity suffers from the removal or damaging of parts is likely to vary in direct proportion to the causal integration of the parts.

The other component of causal relation between Lucy's parts is *causal cohesion*. An entity is causally cohesive to the extent that it behaves as a whole with respect to some process.

> In such a situation, the presence or activity of one part of an entity need not directly affect another, yet all parts of the entity respond uniformly to some specific process (although details of the actual response in different parts of the entity may be different because of the operation of other processes). (Mishler and Brandon 1987, p. 400)

An example of causal cohesion is the reaction of a school of minnows to the presence of a predator. The minnows flare away from the danger as a unit without any significant causal interactions between them. Each is responding on its own to a common external stimulus. That some minnows move away more quickly or slowly does not mean that they are not responding as a unit to the stimulus. Not even relative spatial continuity is necessary for causal cohesion. The parts that form a cohesive unit can be isolated from causal interaction and still be cohesive.

Parts can be related by causal cohesion without being related by causal integration and vice versa. Parts that are isolated from each other to the extent that they are not at all integrated may still be causally cohesive. Some causal phenomena that may initially be taken to be causal integration are actually best described as causal cohesion. Some causally cohesive properties mimic integration when apparent integration turns out to be a common response among the parts to an external stimulus. Lucy's parts are, however, both integrated and cohesive. An example of the causal cohesion between her parts is that they are all affected by a systematic poison. Her parts are cohesive in response to many processes but not with respect to all processes. In later sections, I will discuss the cohesion of biological entities with respect to the forces of natural selection.

Lucy was the product of *sexual reproduction.* Lemurs reproduce sexually and only sexually. Unlike the vast majority of other living things, including many animals, lemurs are unable to reproduce asexually. They neither form single-celled asexual propagules, nor do they split in half to form new lemurs. Lucy originated in the union of a pair of male and female gametes from her parents, which formed a single-celled zygote with a unique genetic constitution – unless she was a monozygotic twin, in which case she would share the same genotype with the other twin. The zygote divides by mitosis. It cleaves into a spherical ball of cells (the blastula), which then goes through a series of invaginations during which it is turned outside in to form a multi-layered gastrula. Different layers of this structure, the ectoderm, mesoderm, and endoderm, form the major components of the lemur body.

There are three general forms of development: somatic embryogenesis, epigenesis, and preformation. During somatic embryogenesis there is no distinct germ line.[5] The same cell lineage can participate in somatic function as a stem-cell lineage and can also give rise to gametes at any stage in the organism's development. All plants and fungi develop through somatic embryogenesis, as do almost all of the multicellular protists and nine animal phyla. The remainder of the animal kingdom develops through epigenetic or preformistic developmental patterns. The germ line is differentiated from somatic lineages in organisms with epigenetic development, but not until rather late in development. Preformistic development is an extreme form of epigenetic development in which "the germ line is terminally differentiated in earliest ontogeny, often under direction of maternal-derived determinants deposited in the egg."[6]

Lemurs develop through preformistic development from a single cell. The cells that are to become the gametes of the next generation are sequestered away very early in development, while the development is still controlled by mRNA from the maternal cytoplasm. Any heritable variant in the cell lineage must occur before the germ line is sequestered. Only germ line cells carry genetic information that can be inherited by the next generation of lemurs. None of the somatic cells can directly pass any genetic information to the next generation of lemurs because they do not have access to the germ line, which is the only way for a variation in a cell lineage to be heritable. The early sequestering of the germ line in the developing lemur makes the whole lemur, rather than any part of it, a *unit of selection.*[7]

3.1 Paradigm Individuals: The Higher Animals

Lucy has an allorecognition system that allows her to distinguish parts of her body from any other living things. Because of its ability to recognize alien life within its boundaries, an animal rejects all grafts not from itself. A higher animal like Lucy can take grafts from other parts of itself and rejects grafts from almost any other source because of the function of the allorecognition system. A skin graft can be transplanted from one part of her body to another part. If skin is taken from one area of an animal and grafted to another area, it grows and spreads outward. If the graft is taken from another animal, the immune system of the graft recipient will kill the graft. It is an immunological reaction that kills the foreign graft, not a lack of circulation or any mechanical problem. "It heals just as soundly, it is as quickly and as richly re-equipped with a working vasculature, and it undergoes the same processes of internal reorganization and repair" (Medawar 1957, p. 154). Exceptions to this rule are discussed below.

Exceptions to the allorecognition rule are important. Monozygotic (identical) twins can exchange skin and other tissues without provoking an immunological reaction. This is because at one point the twins were a single zygote, which split apart later to form distinct beings. An embryo will accept grafts not only from other members of its species but from other species too. Grafts can be exchanged between some special nonmonozygotic twins.

> The non-identical twins between which it is possible to exchange skin homografts are among the most remarkable animals in nature, for they are graft-hybrids or chimeras; each twin is a mixture of cells of two genetic origins, most of its cells being its own, the remainder having been at one time the property of its partner. The exchange of homografts between them later in life does not therefore make them chimeras; it merely makes them more so. (Medawar 1957, p. 150)

This tolerance will extend throughout the life of the organism and will also extend not only to those cells that gained access to the embryo but to any genetically identical cells that enter the organism later in life.

Very few other living things have all of these properties. Using a higher animal as a model of individuality is misleading because the properties it has can be uncoupled from one another. The higher metazoan has the properties most commonly associated with individuality and has them to the greatest extent. The clearest referents of 'biological individual' have been the multicellular higher animals that are most clearly demarcated from their surroundings and from other individuals of the same species.

What makes a living thing an individual? This is not yet an unambiguous question. There are at least ten different criteria of individuality that an adult higher animal exemplifies. These characteristics are not a set of necessary or sufficient conditions for individuality; they are simply a starting point for our investigations.

1. *It is a particular, not a universal or a class.*
2. *It is spatially and temporally localized.*
3. *It is spatially and temporally continuous and composed of heterogeneous causally related parts.*
4. *It would suffer impaired function if some of its parts were removed or damaged.*
5. *It has a single nervous system.*
6. *It rejects all grafts not from itself.*
7. *It is genetically homogeneous.*
8. *It develops from a single cell into a multicellular organism by a characteristic pattern of development.*
9. *It reproduces sexually.*
10. *It is clearly demarcated from other members of the same species.*

The typical higher animal has all of these properties. Most other living things do not. The assumption that there is only one kind of individuality is as unjustified as the inference that an entity that has some of these properties must have all of them.

3.2 OTHER POSSIBLE SOLUTIONS

In the first part of this chapter, I examined the properties of a particular higher animal. My exemplar was a ring-tailed lemur, but the properties I examined were properties common to all higher animals. A whooping crane has these same properties and would have worked just as well. Adult creatures like these are easy to individuate under ordinary circumstances. I did not choose a sponge, an aphid, a strawberry patch, or a slime mold aggregate as an exemplar of individuality because none of these has all of the properties listed above. The fact that these entities do not have all the properties the exemplar has raises an important question. How do we individuate organisms that are not like higher animals? What is an individual slime mold or sponge? Some unit or units of individuation must be developed if there is to be a population biology of such entities. At minimum one must know how to count the number of individuals present at a given time. To count such entities we must learn to individuate them. In this section I explain how to do this.

One approach is to gather together the diverse entities that people call individuals, determine what all of them have in common, and try to discover the

3.2 Other Possible Solutions

necessary and sufficient conditions of biological individuality. J. S. Huxley suggests this approach, though he does not endorse it.

> In determining the nature of Individuality, for instance, we may seek to define it by comparing the different objects we are agreed upon to call individuals and then taking their Highest Common Measure – extracting from them the utmost which is common to all and erecting that as the minimum conception of Individuality. (J. S. Huxley 1912, p. 2)

The problem with this strategy is that biologists, philosophers, and laymen have claimed that an extraordinary range of entities are individuals. A partial list includes single cells, organisms, populations of organisms, symbiotic associations of organisms, species, and higher taxa. These things have very little in common. This approach is not very promising because the common properties do not even differentiate the entities that meet the criteria from many other entities not listed. Nor does it suggest how to individuate apparently paradoxical entities, or give us a means of choosing between different systems of individuation that meet these minimal requirements, and there are many systems that meet them. The prospects for finding a set of necessary and sufficient conditions for being a living individual through the use of this method do not look promising. There appear to be more than one set of criteria for biological individuality in use because of the diversity of the entities that are considered to be individuals. The lack of reasonable necessary and sufficient conditions leads me to believe that there are several distinct concepts of biological individuality.

The difficulty in determining necessary and sufficient conditions may lead one to think of 'individuality' as a family resemblance term like 'game.'[8] Perhaps I should examine what different people call individuals as above, but, instead of trying to find some property common to them all, see how people actually use the term in ordinary language. The term 'individual' when taken to mean whatever people use it to mean will not have a neat set of sufficient conditions attached to it, so perhaps our concept of a living individual need not have necessary and sufficient conditions.

There is nothing wrong with folk concepts unless they are pressed into the service of a technical language as is. My concern is to resolve certain apparent paradoxes and disputes within the context of current biological theory. It would be overly optimistic to think that this could be accomplished by elaborating the details of folk ontology. Because folk notions often include members of distinct natural kinds under a common term, they obscure the very differences I want to uncover (but see Dupré 1993). If we accept the folk notion of biological individuality and explain the extension of 'individual' as a family resemblance term, we may be disguising several more precise concepts under a single term that obscures the distinction between those concepts.

Another way of recognizing that some of the entities to be individuated lack some of the important properties the exemplar individuals have is to claim that a higher animal is an individual to the highest degree, and that other living things are individuals to the degree that they meet the standard set by a higher animal or an idealization of one. J. S. Huxley also suggests this alternative.

> [W]e may search for the movement of individuality through the individuals, and, finding that some are more perfect, some more rudimentary in their individuality, thus establish a direction in which its movement is tending, and from that deduce the properties of the Perfect Individual, possessing then a maximum conception of Individuality. (J. S. Huxley 1912, p. 2)

The bees in a hive are causally integrated, they are just not as integrated to the same degree as the organs in a lemur's body, so they compose an individual of a lesser degree.

Using this method of classifying degrees of individuality, we could begin by placing the higher animal near the top as nearly perfect individuals of the highest degree. We could then classify a living entity as an individual of a higher or lower degree based on how many properties of the paradigm individual that entity has and to what degree it has them. This method has the advantage of recognizing the commonsense individual, the higher animal, as the paradigmatic individual, while still making it possible to apply the same criteria of individuation to an entity that either does not have the same properties or has them to a lesser degree.

These are good reasons for not adopting this method of classification. The first is that it cannot provide a principled ordering of degrees of individuality for a pair of entities that have different subsets of the relevant properties, particularly when those entities have those properties to differing degrees. Does a colony of bees have a higher or lower degree of individuality than a slime mold pseudoplasmodium? Each has a different subset of the properties. I am not sure that there is a principled answer to questions like this one. There are too many variables in the equation to settle on such a system of classification based on degrees.

A second problem with unifying a diverse set of properties under a single measurement of degree is that degree of individuality alone is not a very useful measurement without a closer examination of the particular properties in virtue of which an entity is assigned a degree of individuation. Which properties it has and to what extent are more important than how all of those diverse properties are combined in a unified measure of individuality. A good system of individuation must recognize that particulars can have these properties in different degrees and different combinations. There are ways of recognizing this that do not play some properties against others in a system of ranking.

Another way of recognizing these differences is to adopt a pluralistic account of individuality that applies different standards to different kinds of living entities. A higher animal would be an individual if it had a certain set of properties. A plant, fungus, or lower animal would be an individual if it had a different set of properties sufficient for individuality for that kind of entity, taking into account the prevalence of asexual reproduction, colonial life, and other such properties. Some form of pluralism is called for. But I think that there is a better form of pluralism available. There are several distinct kinds of biological individuality and biological individuals. In the next section I explain what those kinds are.

3.3 THE PROPOSED SOLUTION

The concept of a biological individual as I have developed it so far is a multiple-criterial concept. As we have seen, it is not clear how to classify the majority of living entities because they do not meet all the criteria. None of the alternatives I considered above is adequate. We must reform the concept of an individual in biology so that it is applicable to all living things. To some extent this reformation has already occurred, but it has been obscured by the continued application of the term 'individual' without qualification to several distinct concepts of individuality in biology that share a common, and therefore confusing, term. The properties I listed earlier can be divided into six distinct sets of properties that mark important biological kinds that are often disconnected from one another in nature. The higher animal deserves a special place in biological thought because it marks the coextension of six kinds of individuality. But if we realize that these six concepts can be divided from one another, we can develop a useful vocabulary that describes all living entities, most of which do not embody all of the concepts.

This vocabulary provides a way to make sense of the debate regarding the individuality of biological species, groups of clonal dandelions or aphids, and the discovery of a giant mushroom in Michigan's Upper Peninsula. Many of the current quarrels regarding the unit of individuation for biological entities are rooted in unintentional equivocation between different definitions of individuality. Vacillation between different concepts also clouds the conceptual waters. Fulfilling one concept of individuality does not entail fulfilling all of them. Equivocation and vacillation can be remedied by making these concepts explicit. Doing so will resolve many of the philosophical problems of biological individuation.

Not all questions can be answered by a careful reformation of the vocabulary in which those disagreements are worded. Empirical questions regarding whether a living entity has a property or not cannot be answered by conceptual analysis. Such a reformation will, however, dissolve some disputes and provide a more precise vocabulary in which to state remaining questions so they can be more clearly addressed. Given the variety of different questions someone may be asking through the use of the term 'individual,' the question "Is it an individual?" is ill-formed until we specify what kind of individuality is being investigated. As it is, the question is systematically ambiguous in the same way that the question "How many?" is ambiguous until we specify what exactly we are counting.

We can distinguish at least six different concepts of individuality. The question "Is it an individual?," without further qualification, is ambiguous. When a philosopher or biologist says that a living thing is an individual she may be claiming that one or more of the following is true of that entity.

1. *It is a **particular**. A biological entity is a particular just in case it is neither a universal nor a class.*
2. *It is a **historical entity**. A biological entity is a historical individual if it is composed of spatiotemporally continuous parts.*
3. *It is a **functional individual**. A biological entity is a functional individual if the parts which compose it are causally integrated into a functional unit.*
4. *It is a **genetic individual**. A biological entity is a genetic individual if its parts all share a common genotype.*
5. *It is a **developmental individual**. A biological entity is a developmental individual if it is the product of a developmental process.*
6. *It is a **unit of evolution**. A biological entity is a unit of evolution if it functions as an important unit in an evolutionary process.*

In the remainder of this section I will explain each of these restricted concepts of individuality and why each is important to biology and the philosophy of biology.

'Individual' has been used to describe entities that are *particulars*. This is how P. F. Strawson uses the term in his book *Individuals*.

> For instance in mine, as in most familiar philosophical uses, historical occurrences, material objects, people and their shadows are all particulars; whereas qualities and properties, numbers and species are not. (Strawson 1959, p. 15)

If an entity is a particular it cannot be a universal or a class. All concrete living entities are individuals in this sense of the term. They are particulars. Strawson

3.3 The Proposed Solution

claims that biological species are not particulars, but he was working with an out-of-date conception of taxonomy. Whatever may be the case for other kinds of particulars, any living material particular is spatially and temporally bounded. It is spatially and temporally bounded because the history of life on earth is spatially and temporally bounded. According to the best evidence available, life began on earth approximately 3.1–3.5 billion years ago. All current living things are descendants of the first life. All earthly life or any span of it can rightly be considered a particular. Particularity does not entail spatial or temporal continuity. This criterion of individuality is met by a wide variety of living entities including a cell, a collection of cells, a particular aardvark, or the totality of life on earth.

Particularity has been used as a definition of individuality in the philosophy of biology. Michael Ghiselin (e.g., 1997) argues that biological species are individuals rather than classes. He does so by pointing out that having spatially distinct parts does not make something a class rather than an individual.

> Multiplicity does not suffice to render an entity a mere class. In logic, "individual" is not a synonym for "organism." Rather, it means a particular thing. It can designate systems at various levels of integration. (Ghiselin 1974, p. 536)

As Ghiselin uses the term, 'individual' refers to anything that is not a class or a universal. According to him a biological species meets this criterion. He thinks that a species is a chunk of the genealogical nexus rather than a class of organisms grouped on the basis of shared morphological traits. Most biologists now endorse some version of a phylogenetic species concept or biological species concept that supports this claim. For our present purposes it suffices to explain that it is a concept of a species grounded in genealogical relation rather than resemblance. If Ghiselin is right, each biological species is a spatiotemporally localized historical entity. And if this is true, then a biological species is a particular.

Ghiselin also claims that the conceptual waters were considerably muddied when the notion of an individual (as Strawson uses the term, for example) was imported into evolutionary biology.

> The transfer of concepts across disciplinary boundaries from philosophy to biology proved shallow in the least. In fact, the most elementary of philosophical notions became garbled. 'Organism' was confounded with 'particular' because, at least in part, 'individual' had designated both. (Ghiselin 1974, p. 542)

The garbling to which Ghiselin refers has definitely occurred. The case is even more complicated because other biologists and philosophers have also claimed that biological species are individuals, but they mean by this claim that a species is not only a particular, but that it is causally integrated or causally cohesive and has other properties characteristic of organisms. Some of these claims may be true, but it will take more than proving that species are particulars to prove this. Any entity at any level of biological organization from an organic molecule to a taxon is going to be a particular. This does not imply that each is going to be integrated or function as a unit of evolution or be a scientifically useful unit of classification.

Particularity does not entail the spatiotemporal continuity of the biological entities it picks out. It ensures that the entity is spatiotemporally bounded. A particular can be a scattered object. A historical individual, on the other hand, has spatial and temporal boundaries and is also spatiotemporally continuous. A historical individual must have a single origin and at least potentially a single end. A unicellular organism and all of its descendent cells form a historical individual. A phylogenetic species also has a single beginning, is continuous through time, and has a single end in time, usually when the last organism that is a part of the species dies or when the species splits into distinct lineages. A historical unit is any biological entity that has a single starting point and a single end in time. Many biological entities meet the criteria of historic individuality. Species identified by the method that Ghiselin recommends are one such entity.

As Ghiselin noted above, the term 'individual' has been used to refer to organisms. The terms 'individual' and 'organism' are used almost interchangeably in biology, though they can mean different things. The parts of an entity do not have to be causally integrated for it to be a particular or a historical individual, but they must be integrated if the entity is an organism. The term 'organism' has been taken to mean any functionally integrated living entity.

> We tend to call a biological object an organism if it is spatially separated from others and if its parts are so well integrated that they work only in coordination with others and for the proper function of the whole. (Gould 1984)

'Organism' seems also to be used in a more restricted sense to refer to a causally integrated entity that develops from a single cell by a normal process of development. The difference between these two definitions has often been blurred, but the two definitions do not pick out exactly the same entities. Some entities are functionally integrated but do not develop from a single cell through ontogeny. Instead, through some other process they develop into an adult form that has the same sort of causal integration we find among the metazoan animals. Because these two properties are actually separable, I recommend that we recognize this in our vocabulary of individuation.

3.3 The Proposed Solution

Among the characteristics of the lemur I discussed in the first section of this chapter, a significant subset of these properties concern functional integration. These properties include being composed of heterogeneous causally integrated parts, functioning as a unit of behavior, and typically suffering impaired function if some of its parts are removed or destroyed. These properties can be found in living entities other than individual organisms if 'organism' is used in its more restrictive form. A living entity can be functionally integrated without developing from a single cell. The term 'superorganism' is sometimes used to describe the functional organization of entities composed of more than one organism.[9] The term was necessary because, given the ambiguous nature of 'organism,' there was no term that unambiguously referred only to the functional organization of a living entity independent of other properties such as the origins of the entity.

> From the point of view of human perception, organisms are paradigm individuals. In fact, biologists tend to use the term "organism" and "individual" interchangeably. Thus biologists who wish to indicate the individualistic character of species are reduced to terming them "superorganisms." (Hull 1976, p. 175)

To avoid the confusion which comes from using the terms 'organism' and 'superorganism,' I refer to any sufficiently functionally integrated living entity as a *functional individual*. Functional individuality is determined by the current causal relations between the parts of an entity rather than an entity's developmental history, genetic makeup, or the history of the parts that currently compose it.

An entity can be a functional individual without developing from a single cell according to the normal mode. The development of a colonial siphonophore demonstrates that such a functional individual is possible. "The achievement of siphonophores and chondrophores must be regarded as one of the greatest in the history of evolution. They have created a complicated metazoan body using individual organisms" (E. O. Wilson 1975, p. 386).

A higher animal which has undergone organ transplant surgery is, nonetheless, a functional individual despite the fact that some of its parts were introduced through an unorthodox method. The same is true of chimera formed by grafting a scion to a distinct root stock. The various zooids composing a Portuguese man-of-war form a single functional individual, as do some intensive symbiotic relationships, for example the algae and fungi that together compose a lichen. If Lynn Margulis is right, the eukaryotic cell was itself originally a symbiotic relationship between prokaryotes (Margulis 1981).

There is one problem with the idea of a functional individual. The properties that determine whether something is a functional individual are properties

Individuality and Equivocation

that can be held to varying degrees. There are hierarchies of functionally integrated units in nature. The organelles of a single cell are well integrated, yet that cell may be a part of a multicellular organism in which all of the cells are also integrated into a collective functional individual.

Among higher animals, with some exceptions that I discuss below, the functional individual has a unique genetic constitution. Typically, a higher animal is genetically homogeneous and does not share that genotype with any other functional individual. Even among the higher animals though, this genetically homogeneous unit can be distinct from the functional individual. Identical twins are two functional individuals, but they compose a single genetic individual. A *genetic individual* is the genetically homogeneous unit. T. H. Huxley defined the individual in biology as the total product from a single-celled reproductive propagule. His view was criticized by J. S. Huxley, who thought that the existence of twins provided a reductio ad absurdum of his grandfather's position.

> If anything is an individual on this earth that surely is man; and yet we are asked to believe that though the most of us are true individuals, yet here and there some man who lives and moves and has his being like the rest is none, that he must make shift to share an individuality with another man because the couple happen to be descended from one fertilized egg instead of two. (J. S. Huxley 1912, p. 70)

If T. H. Huxley intended for his definition to correspond to the commonsense notion of an individual, then J. S. Huxley is right. Twins are not ordinarily considered to compose a single individual. T. H. Huxley has, however, identified a unit of individuation that scientists have recently found useful.

This new vocabulary also gives us a method to resolve the debate about the distinction between growth and reproduction from the problem that I explored in Chapter 2. Janzen and Harper both accept that a genetic individual can be composed of spatially isolated functional units. A population of clones is a single genetic individual whether the clones are physically connected or not. They do not agree, however, that the single-celled "bottleneck" of asexual reproduction marks the beginning of a new individual.

Dawkins draws that distinction with the help of a thought experiment in which he considers the case of two aquatic plants, one of which, G, propagates itself through hunks that periodically break off and float away, and R, which propagates itself by manufacturing single-celled propagules either sexually or asexually when it reaches a certain size. Janzen's view entails that unless R reproduces sexually, there is not a significant difference in individuating G and R. G and its asexual offspring form a single individual, as does R and its

3.3 The Proposed Solution

asexually produced offspring. Only if R reproduces sexually is it a distinct individual from its offspring.

Dawkins thinks that the distinction between G and R is important regardless of whether R reproduces sexually or asexually. The important distinction between the two species is "that a lineage of R is capable of evolving complex adaptations in a way that G is not" (1982, p. 257). The difference is that the single-celled stage in either sexual or asexual reproduction provides a new developmental cycle between each generation whether those generations are sexual or not. This single-celled bottleneck allows for major innovations in development if that single cell is a mutant. However, in G's case, although mutations may arise, that mutation cannot result in radical reorganization of the body plan. In light of this example, Dawkins thinks that the individual is best defined as the unit initiated by a new act of reproduction via a single-celled developmental "bottleneck."[10] He ends his discussion of Janzen and Harper on a pluralistic note by remarking that Janzen's distinction between growth and reproduction may be fine for certain ecological or economical questions.

> A sisterhood of aphids may indeed be analogous to a single bear. But for other purposes, for discussing the evolutionary putting together of complex organization, the distinction is crucial. A certain type of ecologist may gain illumination from comparing a field full of dandelions with a single tree. But for other purposes it is important to understand the differences, and to see the single dandelion ramet as analogous to the tree. (Dawkins 1982, p. 259)

I think that there are three important ideas involved in the debate between Janzen and Huxley on one side and Harper and Dawkins on the other. The first is the nature of a genetic individual. Huxley and Janzen believe that there is a significant biological entity that is composed of the entire genetically homogeneous unit. Harper and Dawkins think that there is an important biological unit that is composed of the genetically homogeneous product of a single developmental process regardless of whether that process is initiated from a sexual or asexual propagule.

Both sides of this debate argue that the biological unit that they identify is an important one for understanding natural selection. I think that both of them identify important units and that there is actually not a conflict between the views. Sometimes, as in the two cases that Janzen describes, the best way to make sense of the extensive asexual reproduction is to consider it as analogous to the growth of a spatially continuous plant or animal body. Yet it is certainly true that something important occurs when a new developmental sequence begins and the distinction that Janzen draws does not emphasize development. I do not see the conflict in arguing that both of them identify important units of individuation.

The genetic individual makes no special reference to the developmental processes by which its component parts come into existence. It only notes their genetic characteristics. Sometimes natural selection will act on the genetic individual as a unit. The cases that Janzen describes seem to be good examples of the kind of cases I have in mind. In other cases, the genetic individual does not mark the most important unit. If genetically identical clones from one variety of grape are used as a rootstock for another variety, the disconnected root clones partially compose a genetic individual that does not include the plants grafted onto them. Despite this, they have no chance to reproduce sexually and so have no offspring. Sometimes the unit that Janzen identifies is a useful one; in other situations it is less useful than the one that Harper and Dawkins prefer. I call the unit that Harper and Dawkins identify the developmental individual. Both the genetic individual and the developmental individual are important units because of the relation they have with natural selection.

The term 'individual' has also been used to refer to any entity that functions as a unit or individual with respect to a law. The entity is causally coherent if it responds to some pressures as a unit or single entity. In her description of the nature of species, Mary B. Williams points out that the notion of an individual in this sense should be detached from our perceptual abilities or prejudices based on them. She gives the example of an atom-sized, very short-lived scientist inside a moving baseball. Because of his size and duration he does not perceive the baseball as a unified entity. She then draws the moral of the story.

> Our microscientist sees the baseball as a set of molecules, just as we see a species as a set of organisms. But with respect to Newtonian laws, the baseball is an individual – that is, because of the cohesive forces that hold its molecules together, it acts as a unit with respect to these laws. The claim that species are individuals in evolutionary biology is a claim that species are held together by cohesive forces (e.g., common selection forces on a common gene pool) so that they act as units with respect to the laws of evolution. (Williams 1992, p. 322)

The most important form of cohesion for biological entities is cohesion with respect to the forces of natural selection. For the most part, the parts of a higher animal face selection as a unit because the germ-line cells are sequestered and the somatic cells cannot bud off to form a new multicellular organism. Other entities that are not higher animals can also to some extent react cohesively to selective pressure.

Lewontin points out that natural selection requires phenotypic variation, differential fitness (rates of survival and reproduction), and heritability of

3.3 The Proposed Solution

this fitness. Although Darwin may have originally conceived of the notion of evolution by means of natural selection as a process primarily involving organisms, Lewontin notes that these principles can be generalized to other kinds of entities (Lewontin 1970). Any entity at any level of biological organization that meets these criteria is a "unit of selection." Different biologists have argued that a wide range of entities at different levels of biological organization – genes, chromosomes, cell lineages, parts of organisms, colonies of eusocial insects, populations, species, and higher taxa – all may have, under some circumstances, the necessary properties to face the forces of natural selection as cohesive units. For a fuller exploration of these topics see Brandon and Burian (1984).

On which of these diverse biological entities does selection operate? This question can be divided into two more specific questions regarding the role of those entities. Is the entity a replicator? Is the entity an interactor? A replicator is a living entity that reproduces itself directly with a high degree of fidelity. An interactor (or vehicle) is "an entity that directly interacts as a cohesive whole with its environment such that replication is differential" (Hull 1980, p. 318). In organism-level selection, genes are replicators and the body is the interactor.

Replicator and interactor are functional roles that could be filled by entities other than genes and organisms. There may be replicators other than DNA and interactors other than phenotypes. There appear to be a variety of suborganismic-level entities – for example, cell lineages – that sometimes function as interactors. The case for higher levels of interactors is more contentious. The conditions required for a higher-level entity, such as a herd, to function as an interactor are clear, but it has not been proven that there are actual group-selection processes that merit treating groups of organisms as interactors. Similarly, it is a contentious issue in biology whether there are replicators other than lengths of DNA (Sterelny, Smith, and Dickison 1996; Griffiths and Gray 1997). Richard Dawkins thinks that all evolutionary developments are ultimately for the benefit of "selfish genes," but though the conditions for being a replicator are fairly clear, whether any other entity can function as a replicator is empirically undecided. Replicator and interactor are functional roles that may be occupied by many different kinds of entities that have little else in common. Though I think that it is important to note that selection may act on entities other than the organism, I do not think that there is a kind of individuality characterized by playing either of these roles.

The kinds of individuality I identify in the previous section are important kinds for living entities. These kinds can explain the diverse nature of real cases with which biologists contend. This set of kinds is not as simple as the folk notion, but it is no more complex than the phenomena it describes.

Biologists need to be able to count living entities of various sorts unambiguously, and the commonsense notion of individuality is inadequate for the task. Harper, more than anyone else, seemed to realize the virtues of introducing more than one level of population structure to a theory of individuality rather than trying to construct a univocal concept. My theory is an expansion of his view.

My findings in this chapter explain why philosophers who write about individuation must pay attention to real issues in biology. The fact that individuality *simpliciter* does not work for biological entities has important implications for a philosophical theory of individuation. Individuals of these different kinds tend to overlap with one another; an adequate theory of individuation must provide a consistent account of this overlap.[11] The kinds I identified have different persistence conditions, so it makes no sense to identify an entity individuated by one of the concepts with an entity individuated by another, even if they are entirely coextensive, as they often are among higher animals. Considering a wide range of biological phenomena forces us not only to expand our concept of biological individuality, but to recognize that the normal cases that we began with are actually more complex than we initially thought.

In the next chapter I argue that some features of a living entity's biological origin are essential to it and explore the implications of this view for an entity's sex and species membership. In Chapter 5 I return to the kinds that I identified in this chapter and explain how individuals of those kinds persist through time.

4

The Necessity of Biological Origin and Substantial Kinds

4.1 A VALID ARGUMENT FOR SORTAL ESSENTIALISM

In this chapter I argue that a living individual of any kind necessarily has its actual biological origins. This assertion may initially seem to be a scholastic aside to the question of individuation, but a short summary of the previous chapters will show that the necessity of origin will play an important part in my overall argument. I have demonstrated that the substantial kinds philosophers have generally considered to be the substantial kinds for living individuals do not seem appropriate when applied beyond the most basic familiar examples. I have proposed an alternative to these substantial kinds that better explains biological individuality and grounded these suggestions in a theory of natural kinds.

I established that the existence of real living individuals depends on the existence of real substantial kinds that those individuals exemplify. The substantial kind for a living thing marks the boundaries of the sort of change that entity can undergo without ceasing to exist. If an entity is a living thing of a substantial (natural) kind, it cannot cease to have the properties characteristic of that kind without ceasing to exist. An entity's substantial kind marks those properties as essential to the continued existence of a thing if it originates as a member of that substantial kind. If x is of kind K, x cannot cease to have the properties essential to kind K.

As it stands, this argument demonstrates only that there are some changes that the entity cannot undergo and survive. It does not prove that the entity could not have existed *without ever being a member of that substantial kind*. Additional premises are necessary to support the conclusion that a living entity is necessarily a member of the substantial kind of which it is actually a member.

Penelope Mackie has attacked the inference from the individuation of entities through a substantial sortal to sortal essentialism (Mackie 1994). Sortal essentialism is the view that if an entity is actually individuated by a substantial kind in the actual world, it necessarily is individuated by that sortal. She illustrates this inference with an example. Aristotle belongs to the substantial kind *man*. *Man* and *centipede* are incompatible sortals. Because Aristotle is a man, there are other properties he cannot have – for example, he cannot be a *centipede*. Being a centipede precludes the possibility of *becoming* a man, and vice versa. This is true because the substantial kinds *man* and *centipede* have incompatible existence and persistence conditions.[1] The same entity cannot be simultaneously both a man and a centipede, nor could a creature change from a man into a centipede without ceasing to exist. Therefore, because Aristotle is a man and *man* and *centipede* are incompatible substantial sortals, Aristotle could not have been a centipede. Mackie challenges the validity of sortal essentialism:

(P1) Aristotle is a man.
(P2) Belonging to the substantial kind *man* is incompatible with belonging to the substantial kind *centipede*.
(P3) Aristotle cannot change from being a man to being a centipede.

(C) Therefore Aristotle could not have been a centipede.

She is right to challenge this argument. These premises do not entail that Aristotle could not have been a centipede for his entire life. Without further premises we cannot prove that Aristotle could not always have been a centipede, a conclusion quite different from the assertion that if Aristotle is a man, he cannot change into a centipede without ceasing to exist.

While Mackie disavows sortal essentialism, others such as Wiggins defend it because it provides the principle of individuation for individuals of that kind.

Essentiality of Principle of Individuation (EPI): If an individual x has a principle of individuation P, then x could not have existed without having P.

He reasons that Aristotle could only have been a centipede if he had a principle of individuation different from his actual one. Because Wiggins thinks that this is impossible, it is also impossible for Aristotle to have been a centipede.

Mackie argues that EPI either relies on an unwarranted principle or is vacuous. In either case, it is not to be confused with a principle about temporal essentialism (AIP).

4.1 A Valid Argument for Sortal Essentialism

Absolute Identity Principle (AIP): An individual cannot change its principle of individuation over time, nor can it have two different principles of individuation simultaneously.

Mackie does not object to this principle, but she does challenge an interpretation of EPI.

EPI (1): If an individual x has a principle of *distinction and persistence* P, then x could not have existed without having P.

Mackie thinks that there are no good arguments for EPI (1). She considers a series of arguments based on the absolute principle of identity and finds them wanting. She rejects arguments that depend on inferring ways an entity could or could not have been from the kinds of changes an entity could or could not survive. She thinks that these arguments fail because of the logical gap between changes a thing could endure and counterfactual claims of what the thing could have been from the start. Any attempt to infer what a thing must be from the changes it can endure needs to be bolstered by further argument.

One reason that she considers and rejects is based on the apparent absurdity of a thing having a different individuative principle. "The idea that a cat could have been a butterfly is ridiculous. What could make it the same individual if it is as different as that?" (Mackie 1994, p. 328). As Mackie notes, this apparent ridiculousness is not an argument against the position that such a change is possible.

I think, however, that an argument can be developed to explain why such a change is ridiculous. We can add premises to the argument Mackie criticized as invalid. One way to bridge the gap in the argument is to demonstrate that some aspect of the actual origin of a living thing is essential to it. If we insert these premises, we get the following argument, which is a generalization of Mackie's argument about Aristotle:

(P1) Aristotle is a man.
(P2) Aristotle necessarily has the origins he actually has.
(P3) If Aristotle has his origins necessarily, then he is essentially of the substantial kind that he actually is.
(P4) If a thing is a member of a substantial kind, then it cannot be a member of another substantial kind with different persistence conditions.
(P5) *Centipede* and *man* have different persistence conditions.

(C) Aristotle is not and could not have been a centipede.

This argument is valid. Its soundness depends on the truth of the additional premises. (P1) is true. I have argued for (P4) and (P5) in Chapter 1. In section 4.2 of this chapter I will argue for (P2) and (P3).

4.2 THE NECESSITY OF BIOLOGICAL ORIGIN

Is the actual origin of a living thing necessary to it or is it a contingent fact about that entity? First, I should clarify the sense of origin with which I am concerned. By having the same origins I mean coming from the same parent in the case of an asexually reproducing entity or from the same parents in the case of a sexually reproducing entity. This is not quite specific enough because it would leave open the possibility that I could be one of my sisters in another possible world or that an asexually produced entity could be any of the other vegetatively produced offspring of the parent. For a sexually produced entity the sense of origins that I am most concerned with is that the entity come from the sperm and egg that it actually came from. And that in the case of an asexual entity that it necessarily comes from the asexual propagule, an egg produced by parthenogenesis, that produces it. The necessity of origin for living things then will be stated as

The Necessity of Origin Thesis: A living entity that has an origin in some possible world has the same biological origins in every possible world in which it exists.

There are intuitions that seem to support the necessity of origin, but there are also intuitions that support the contingency of origin. I feel as if I can imagine myself living in ancient Babylon born of ancient Babylonian parents while still being me. If an organism's actual origin is essential to it, then the situation I have imagined is impossible. The opposing intuition is strongest when we think of living things other than people. It is unclear how we could in principle identify a particular petunia in the actual world with a petunia in a possible world unless it came from the same gametes from which it developed in the actual world.

I do not think there is a good argument to prove the incoherence of rejecting the origins of a living thing as essential to that thing. A coherent rejection of the necessity of origins can be constructed. But such a rejection would entail accepting principles that seem to be wrong. I will argue that a living thing's origin is essential to it because of what the rejection of this principle entails.

I present two kinds of arguments for the necessity of biological origins. The first is based on the apparent impossibility of identifying an entity in a

4.2 The Necessity of Biological Origin

possible world without its actual origin to identify it as the same entity (Prior 1960). In this argument the necessity of identity thesis above is a particular application of a more general account of transworld identity. The second line of argument advocates the necessity of origin because of the arbitrary nature of the transworld identities compatible with its rejection. If a living thing can develop from a different propagule in that world, it may be compossible with the propagule from which the entity originates in the actual world (Forbes 1980, McGinn 1976). I will explain Prior's view first.

Prior assumes that there must be a factual basis for transworld identity. If a statement about transworld identity is true, it must be true in virtue of some further state of affairs. It cannot be a brute fact that a living thing in the actual world is identical with something in another possible world. This is not an epistemological assumption. It does not presume that anyone can make any particular transworld identification, only that principled reasons for identifying an entity in the actual world with an entity in a possible world do exist. Otherwise, what makes one such identification true rather than another?

He then examines what those factual conditions must be for an actual entity to be identified with an entity in a possible world. He uses Julius Caesar as an example. Which worlds contain Caesar? There are good reasons for thinking that no world other than the actual one contains Caesar. Imagining an entity in the possible world with many of his attributes is insufficient to identify that entity as Caesar because he is "not a property or collection of properties." Caesar was a real living human being. Under what conditions could he be identical with someone in a merely possible world?

Toward answering this question, Prior asks what makes a possible world.

> We might say that a possible world is (i) one of the alternative possible future outcomes of the present actual state of affairs; or by natural extension (ii) anything that was a possible state of affairs in the preceding sense i.e. an outcome of some past state of affairs which was possible at the time, though it may by now have been excluded by what has taken place actually instead; or finally, (iii) we may use the phrase for anything that constitutes a 'possible world' in sense (i) or (ii), together with its past, so that a possible world in this last sense is a total course of events which either is now possible or was possible once. (Prior 1960, p. 69)

All type (i) possible worlds contain Caesar because they deviate from the actual world only after Caesar's beginning and end of existence. Some type (ii) possible worlds contain Julius Caesar. What standard can we use to determine which of these worlds actually contain Caesar?

The type (ii) worlds that differ from the actual world only after Caesar's conception contain him. What about the possible worlds that differ from the actual world before he was conceived? Could Caesar have had different biological parents? If so, when would that world deviate from the actual world? If it deviates after Caesar is conceived, it is too late for Caesar to have had different parents. If it is possible for Caesar to have had different parents, the world must deviate in such a way that Caesar has a different origin.

Prior thinks that no possible world that deviates from the actual world in this respect contains Caesar because before he was conceived there was no individual identifiable as Caesar to be the subject of the possibility of being born to different parents.

> Julius Caesar, i.e. a certain now-identifiable individual, did at a certain time begin to exist. But before that time, the possible outcomes of what was going on did not include the starting-to-exist of *this* individual. However, they did include the possibility that there should be an individual born to these parents, who would be called 'Caesar', would be murdered on the Ides of March, and so on; and this possibility was in fact realized when Caesar was born and underwent all these things. (Prior 1960, pp. 70–71)

Before Caesar was conceived he was not an identifiable individual subject to the possibility of being born to his parents or any others. It was possible that someone with his attributes would be born, but he was not yet there to be the subject of that possibility.

> The possibility that an individual should begin to exist and doing and undergoing the things in question will constitute a realization of this possibility; yet one cannot say of any individual that what was possible was that *he* should begin to exist and do and undergo these things; there just cannot be a possibility of that sort except with respect to what already exists. (Prior 1960, p. 71)

According to Prior, Caesar is not there to be a real subject of our speculation until he begins to exist. We would have no means of specifying that person specifically. If there is no principled way for anyone to identify an entity in a counterfactual situation, then that entity cannot be said to exist in that counterfactual situation unless counterfactual identity is based in brute fact, a possibility that Prior rejects. If Prior is right, then Caesar could not have been conceived by different parents, or by his actual parents in any way other than the way in which he was actually conceived. Thus, he must have his actual biological origins.

Prior makes the most plausible case for his conservative view on transworld identity by presenting it at its strongest point, that the living thing could not have had different parents. His argument supports the stronger claim that every aspect of Caesar's origin is essential to his existence. Not only could Caesar not have had different parents, it seems as if there are type (ii) worlds

4.2 The Necessity of Biological Origin

that do not contain Caesar because they deviate from the actual world in more mundane ways before his conception. If there were any difference in the circumstances previous to the time of Caesar's birth, its timing or location in that possible world would preclude the identification of any entity in that world with Caesar, because Prior thinks that there is no clear way to identify entities such as Caesar in that world.

If qualitative identity is insufficient, something else must link entities in that possible world with entities in the actual world. Though none of the other circumstances of his origin seems to be as important to this existence as Caesar as the biological features of his origin that Prior concentrates on, these are still entailed by his argument. Sentences such as "If the taxi had not come in time she would have given birth to you on the sofa bed" are necessarily false because there is no way except for complete agreement in past circumstances to connect entities in one world with entities in another. This kind of possibility does seem to be real.

I think that where Prior's argument goes wrong is in the inferential jump from the plausible idea that transworld identity must have a factual basis to the view that if a biological entity does not have exactly the same origins as it actually had, then there can be no factual basis for thinking that an entity in that possible world is identical with the entity in the actual world. Biological origins, coming from the same propagules that one actually came from, provide just such an anchor for assertions of transworld identity and do not require that the possible world in question not deviate at all from the actual world before the entity begins to exist, so long as the world has features to fix the transworld identity of the gametes from which the entity develops.

Both Graeme Forbes (1980) and Colin McGinn (1976) argue that an individual's biological origins are essential to it. Their arguments are distinguished from those presented by Prior by the special emphasis they place on the biological processes involved in the development of a living thing rather than general issues about possible worlds. With a minor reservation about the relation between an organism and the zygote from which it develops I agree with the position that they develop.

Forbes starts his argument with two assumptions.

1. Identity is a necessary relation.
2. Transworld identity claims cannot merely be true, there must be something intelligible that makes them true.

To illustrate his second assumption, Forbes asks whether a person (A) who is not an identical twin could have been an identical twin. Identical twins occur when a single-celled zygote splits into a pair of cells with identical genetic content, each of which contains half the genetic material of the original cell. Instead of remaining stuck together as in the development of a non-twin, the cells separate, and each develops into a fetus. Could the zygote from which A developed have split to form twins B and C? If so, is A identical with either B or C? Forbes says A could not be either B or C because A could not be identical with B and also identical with C (unless $B = C$), and there is no intelligible reason to choose one rather than the other. What he actually shows is that the zygote could not be identical to both B and C. He does not consider the possibility that the zygote, though an essential part in the development of A, is not identical with A. Perhaps another example would be better.

An oak tree develops from an acorn. Forbes calls the acorn the oak tree's generation system. He sets up a thought experiment to show the implausibility of rejecting the necessity of origin for a living entity. Imagine an oak tree in the actual world that came from a generation system S. For example, S could be an acorn that Teddy Roosevelt planted on his Long Island estate. It has now grown into a gigantic oak tree. Someone skeptical about the necessity of origin must accept that there is a possible world, w, in which that same oak tree exists, but that in that world instead of developing from S, the acorn that Roosevelt planted, it develops from S', a distinct acorn.

S' is compossible (both could exist in the same possible world) with the presence of S, and both can grow into oak trees. What this means is that there is another possible world, w', in which Teddy Roosevelt plants acorn S and it grows into an oak tree. Somewhere else, a distinct acorn S' also grows into an oak tree. Which tree in w' is identical with the actual tree on the Roosevelt estate in this possible world that contains S, S', and the two resulting oak trees?

Someone skeptical about the necessity of biological origin must answer S, S', or neither. Forbes claims that none of these options is acceptable.

> All three possible answers have the consequence that a generational system which has a given output in one world w has a distinct output in another world w', where the salient features of w' can be described by simply extending the description of the salient features of w, so that w' contains all the relevant features of w. (Forbes 1980, p. 356)

If the skeptic answers either that the output of S in w' is identical with the actual tree in Roosevelt's yard in this world, or that neither tree in w' is identical with that actual tree, then he needs to explain why the presence of S, the actual generation system, has this effect on the transworld identity between S' in w' and the oak tree that Roosevelt actually planted in the actual world. The world w that the skeptic described where only S' occurs and the

4.2 The Necessity of Biological Origin

tree that develops from it is identical with the tree that Roosevelt planted does not differ from w' in any way that should affect S'. The only relevant difference between w and w' is that S is in w'.

If the skeptic says that the tree that develops from S' in w' is identical with the actual tree that came from the acorn that Roosevelt planted, then Forbes says that the skeptic should consider a third possible world, w''', in which only S (as it occurs in a world like w' that contains both S and S') is present. If there is no principled reason why S is not the generation system of a tree identical with the actual tree, and if it is identical with the actual tree in world w'', then there appears to be no good reason why it was not the actual tree in w' where both S and S' occurred.

Though it is implausible, the skeptic could claim that the nonoccurrence of S somehow affects whether or not the tree that develops from S' in that possible world is identical with the actual tree. If the skeptic insists that this is an essential feature of such a world, then he must give an account of how what goes on somewhere else affects whether a tree in that possible world is or is not identical with the actual tree. Any such principle would make identity an extrinsically determined relation.

> These arguments against the skeptic all exploit the fact that he proffered alternative generation systems for actual objects which were compossible with their actual generation systems, forcing ad hoc variations in the identity of certain objects from world to world according to whether or not the actual generation system also occurs in this world. (Forbes 1980, p. 359)

When we follow the history of a living thing through time, we reject the principle that causally unrelated extrinsic factors can determine whether or not it is a genuine continuant. We should also reject the related modal principle implied by the thesis that the relation between a living entity and its generational system is a contingent one.

McGinn also attempts to support the necessity of biological origin by exploring the odd results of allowing interruptions in the continuity of a living thing's actual ancestry and development. He considers three relations, that of the zygote to the adult organism, the gametes to the zygote, and the biological parents to gametes. If each of these three relations is a necessary one and this necessity is transitive, he thinks that he can prove that a living thing's origins are necessary to it. He claims to need two assumptions for his argument.

i. An adult organism is identical with the zygote from which it comes.
ii. Transtemporal identity is necessary.

Biological Origin and Substantial Kinds

If a living entity is identical with the zygote from which it develops and transtemporal identity is necessary, then a living entity necessarily comes from the zygote from which it actually comes. There is a problem with this view as McGinn presents it. Claim (i) seems to be false. If his argument depends on (i) it will not work. McGinn says in support of (i) that

> Adults are commonly identical with children, and children with infants, infants with fetuses, and fetuses with zygotes. Any attempt to break the obvious biological continuity here would be arbitrary. (McGinn 1976, p. 132)

I disagree with McGinn. I think that there is an obvious, nonarbitrary break. Adults, children, infants, and fetuses all share properties that a zygote lacks. All of them are highly organized multicellular entities. None of them is prone to splitting into multiple entities. A zygote is the single cell formed by the fusion of a sperm and egg, which can split to form twins. Among fissiparous animals, the zygote may become even more fragmented. Further, there is a reductio that shows that (i) cannot be true. Identical twins develop from a single zygote. Despite the name, identical twins are not identical. If each is identical with the same zygote from which it develops, by the transitivity of identity, the identical twins would actually have to be identical. I will temporarily leave aside my doubts about the identity relation holding between the zygote and the resulting organism. Ultimately McGinn does not need to rely on premise (i).

He recognizes that even if an organism is identical with the zygote from which it comes, the same cannot be true of the relation between the two gametes and the resulting organism. A pair of gametes cannot be identical with an organism because there are two gametes and one living entity, and identity is a one-one relation. If it were possible for an organism to come from different gametes, those gametes might be compossible with the organism's actual gametes. If the gamete-zygote relation is contingent, then it is possible that the living entity could have come from another pair of gametes even if its actual gametes were also in that world. In that world, the organism that develops from the different gametes, rather than the organism that develops from the actual gametes, is identical with the organism in the actual world.

The causal and spatiotemporal continuity between the gametes that actually form a living entity and that entity favor the necessity of having the same gametes.

> Just as you must have come from the zygote you came from because you are diachronically and developmentally continuous with it, so you must have come from the gametes you came from because you are similarly continuous with them. I shall call this relation *d-continuity*. The intuitive content of the idea of

4.2 The Necessity of Biological Origin

d-continuity is given by the concept of one thing or things *coming from* another thing or things. (McGinn 1976, p. 133)

Unlike identity, d-continuity does not require the persistence of the things that develop. All biological fusion and fission relations give rise to a necessity of origin. The argument in the form he presents it does not actually depend on an adult being identical with the zygote from which it developed; it is enough that the adult is d-continuous with the zygote, which it is. D-continuity confers rigidity on the relations involved, thus accounting for the necessity of origin. The biological processes leading from a parent to an offspring are organized and causally unified processes. I find d-continuity to be a plausible principle, but McGinn makes no argument for it beyond the intuitions it formalizes. The real argument in favor of the position on identity that McGinn and Forbes support comes from seeing what the rejection of the principle entails.

The arguments I have considered above do not conclusively prove that someone who denies the necessity of biological origin must fall into contradiction. To avoid contradiction, someone doubtful of the necessity of origin will have to reject at least one extremely plausible principle underlying the arguments that Prior, Forbes, and McGinn present. Along with the necessity of origin, the doubter will have to reject the view that transworld identity is necessary, that there must be some factual basis for transworld identity or that transworld identity is an intrinsically rather than extrinsically determined relationship. I take these arguments to show that we have good reason to believe (P2).

(P2) Aristotle necessarily has the origins he actually has.

At minimum, Aristotle could not have developed from a different zygote. He could not have had his origins in a different sperm or egg. Because the sperm and egg themselves are also biological entities subject to the same constraints as Aristotle, this entails that he could not have had different parents.

This leaves (P3) to be explained.

(P3) If Aristotle has his biological origins necessarily, then he is essentially of the substantial kind that he actually is.

A living entity's genetic makeup is determined by the genetic makeup of the zygote from which it develops, mutations excepted. Because it must develop from just the biological origins it actually has, it must start with the genetic makeup with which it actually starts.[2] The entity's entire ancestry is also determined by the necessity of origin. My argument turns on the connection between origin and substantial kinds. A living entity has the properties it has as the result of the interactions between its genotype and its environment. Because each living thing necessarily has its actual genotype, it has that genotype in every possible world in which it exists. So the only variable that could be introduced would be in the environmental factors involved in development. The vast majority of phenotypic properties, such as having blue eyes, having eyes, or having four limbs, are all partially determined by environmental factors distinct from genetic ones. If these properties are essential for belonging to a particular substantial kind, then there will be no way to be certain that an entity must be of that substantial kind based only on its ancestry.

By the definition of substantial kind, if an entity has a beginning in time, it must be a member of whatever substantial kind it belongs to from the moment it begins to exist. The substantial kind sets the persistence conditions for individuals of that kind. If an entity does not have a beginning in time, the substantial kind still identifies its persistence conditions through the sorts of changes that it cannot survive. In another possible world there is every reason to believe that the same initial conditions would also result in that individual being of the same substantial kind.

The only way that the same origins would not result in an individual of the same substantial kind is if the substantial kind is determined by factors extrinsic to the biological processes that are occurring at the entity's origin. Because we have already determined that the nature of a substantial kind is determined by the properties of the individuals that are members of that kind and not determined by external factors, the necessity of origin entails that all living entities necessarily belong to the substantial kind to which they actually belong, so (P3) is true and a good argument for sortal essentialism has been established. In the next two sections I explore the implications of the necessity of origin on the sex and species of an entity.

4.3 SEX

The necessity of biological origin entails that an entity must initially have the same genotype it has in the actual world, though over time it may deviate from its genetic origins. There is often, though not always, a close connection between the sex of an organism and its genetic structure.[3] Does the necessity of origin make the sex of an organism a necessary property of that organism? The existence of sequential hermaphrodites is sufficient at least to prove that it is not always so. The answer to this question depends in part on the extent to which genetic factors determine the sex of an organism. Not all living things reproduce sexually. For those that do, the genetic basis for the sex

4.3 Sex

of that entity is a necessary feature of it because it is included in the overall genetic structure, which is a necessary feature of the organism at its origin. This genetic structure should not be confused with the physical traits typical for either sex. The phenotype typical of either sex is dependent on variable environmental and hormonal factors in addition to genetic ones. Among diploid organisms, the (genetic) sex of an individual is determined by a single pair of sex chromosomes. Human females have two X chromosomes, males have an X chromosome and a Y chromosome. Sons get their X chromosome from their mother and their Y chromosome from their father. If all sexual organisms shared this property, it would be easier to argue for the position that the sex of an organism is determined by its genetic makeup, which will, of course, include the sex chromosomes. By this reasoning, if an organism has two X chromosomes it is a female. If it has an X and a Y chromosome it is a male. But things are not this simple.

Some insects, such as grasshoppers, have no Y chromosomes. Female grasshoppers have XX sex chromosomes. Male grasshoppers do not have Y chromosomes. The structure of a male grasshopper's sex chromosomes is represented as XO. The 'O' is not a chromosome; instead it marks the absence of a second sex chromosome. Male birds, moths, and butterflies have two X chromosomes. The female of these species has only one X chromosome. She may or may not also have a Y chromosome.[4] These facts make it difficult to generalize about the link between sex and origins beyond the inference that to the extent that sex is determined by genetic factors, it is a necessary feature of the individual. If it were only a matter of qualifying our remarks about sex and genetic structure, it might be worthwhile to add those qualifications and then argue that the sex of an organism is essential to it, but that what constitutes being a member of a sex may vary for different kinds of organisms. The case is complicated further by the fact that environmental sex determination occurs in many species in which obviously contingent environmental features trigger sex changes.

There are still further facts about development that make the attempt to find essential links between phenotypic sexual traits and the genetic makeup of the individual futile. Up through about the first month of development, human embryos, whether male or female, have a set of female (Müllerian) ducts and a set of male (Wolffian) ducts. At this point the genetic sexes typically diverge and develop sex-particular traits. When there are developmental abnormalities the connection between the genetic basis of sex and the phenotypes typical to them breaks down. Some embryos with XY sex chromosomes lack the enzyme that converts testosterone to androgen dihydrotestosterone (DHT) during fetal development. Babies without this enzyme are born with ambiguous or characteristically female external genitalia and many are raised

Biological Origin and Substantial Kinds

as girls. However, these babies have the internal genitalia of a male. During puberty, testosterone rather than DHT induces physiological changes in males. Pubescent development is relatively unaffected by the inability to synthesize DHT. The results are surprising. During puberty, an adult anatomical male develops from what was apparently a prepubescent girl with XY sex chromosomes. This condition is common enough in some areas in the Dominican Republic to have received a colloquial name, "Penis-at-Twelve."

There are similar cases in which apparent physiological males have turned out to be females.

> An apparently normal boy baby has convulsions a few weeks after birth. Medical workups show that he has a disease that interferes with the adrenal glands' synthesis of the hormone cortisone. Further examination reveals that this "boy" has two X chromosomes, ovaries, oviducts, and a uterus. In fact, "he" is really a girl whose vaginal lips have fused to become a scrotum and whose clitoris has developed into a penis. (Ehrhardt and Meyer-Bahlburg 1981)

Anatomical differences are not determined by the genetic structure that underlies them because of the plasticity of those anatomical features in response to environmental variation. Though the typical XX human has typical female anatomy and the typical XY human has typical male anatomy, the genetic structure of the individual does not guarantee that this will be the case. The normal developmental sequence cannot be isolated from the contingent environmental conditions under which it develops. These conditions can be changed after the entity begins to exist in ways that will alter the anatomy of the organism. The sex of an organism, if we characterize that sex as the phenotypic properties characteristic of either sex, is not an essential property of that organism and cannot be divined from the necessity of origin alone.

4.4 SPECIES MEMBERSHIP AND THE NECESSITY OF GENEALOGY

The necessity of origin entails that a living entity necessarily has exactly the ancestry that it actually has. For example, consider the origins of a particular English setter puppy. By the necessity of biological origin, if that dog developed from gametes, then it must have developed from the actual gametes from which it actually developed. For each of those gametes to be the gametes they are, they must have been produced by the same dogs that actually produced them. For this to be the case, the dogs that produced those gametes, the parents, must also be the same entities they actually are. For each parent to be what he or she actually is, that parent must itself have come from the same gametes and therefore the same parents from which it actually came. No

4.4 Species Membership and Genealogy

entity that is a link in the genealogical progression leading to the conception of that particular pup could have been a different biological entity than it was, if the pup in question was to exist. What does the necessity of genealogy tell us about species membership? The answer to the question depends on the principles that govern the organization of organisms into species, and determining just what these principles are is an unsettled issue in philosophy and taxonomy.

It is a common and mistaken notion that an organism is a member of a species of which it is a member just because of morphological similarity between that entity and other members of the same species. The early Linnaeus believed that the organisms that he grouped into species were more or less identical with their ancestors. Those ancestors were themselves the result of divine creation. Animals of the same species shared a common fixed essence with each other and their ancestors. The assumption that species are kinds based on morphology has been hard to escape.

Under this conception of species, a particular species is the class composed of any and all organisms that have the morphological traits that define that specific essence. For example, a plant is a member of the species *Toxicodendron radicans* just because it has important physical properties in common with the other plants we classify as poison ivy. If this were actually the case, the plants that we call poison ivy would share the essential morphological properties of an exemplar or archetype poison ivy plant. A plant could not be poison ivy if it lacked these properties and it would necessarily be poison ivy if it had them.

This is not how organisms are grouped into species. There has been a taxonomic revolution to square taxonomical methods with the historical nature of evolution by natural selection. Pre-Darwinian taxonomy was primarily concerned with morphological similarity. Darwinian evolution emphasizes genealogy in taxonomic organization and homology over homoplasy. Taxonomists have proposed using considerations of overall morphological similarity (numerical phenetics) or the links of actual or potential interbreeding and reproductive isolation (the biological species concept) to provide objective criteria for determining species membership. In most cases, properties such as potential for interbreeding or overall similarity are used as the final arbiter of genealogical relations. There is still a controversial tension between using morphological and strictly genealogical criteria in parsing organisms into species, but most taxonomists have now adopted some version of the biological species concept or a phylogenetic species concept. Neither is based on the morphological similarity of the organisms involved.

This movement has been accompanied by the claim that species are individuals rather than classes. As I explained in Chapter 3, the notion of individuality used in contemporary evolutionary biology stands ambiguously among several more precise concepts. The least controversial interpretation of the idea that species are individuals is that species represent genealogical units and that species membership is to be determined through relation to a common ancestor. A more controversial claim is that species display other kinds of individuality, particularly a sort of causal cohesiveness or integration by which they are unified by gene flow and barriers to gene flow and so forth. Because of this, a species is a kind of historical unit, and as such, it is not to be identified with a spatiotemporally unrestricted class of organisms sharing a morphological essence. In a genealogically based taxonomy, a creature from outer space cannot be classified as a tiger even if it is morphologically identical with an actual tiger because it is not a descendant of tiger parents. It does not bear the right genealogical relation to other tigers, so it cannot be a tiger.

There are two aspects to a taxonomic theory, grouping criteria and ranking criteria. These criteria may be quite different from one another. A phylogenetic taxonomy groups organisms by evolutionary genealogy. Grouping criteria set the standard for collecting organisms together into a group. Ranking is concerned with placing that group within a taxonomic framework. Is it a variety, a subspecies, a race, a species, a genus, a kingdom?

A genealogically based taxonomy groups organisms into monophyletic lineages. A *monophyletic* lineage consists of all and only descendants of a common ancestral population. Monophyletic groupings sometimes result in unconventional divisions of species. Traditionally crocodiles are grouped together with lizards, snakes, and turtles in the class Reptilia, and the birds are in their own class, Aves, based on the morphological similarities between lizards, snakes, turtles, and crocodiles, and their common morphological differences from birds. By monophyletic criteria, crocodiles are grouped with the birds, not with the turtles, snakes, and lizards, which are grouped together, so morphology does not play a primary role in the grouping of organisms.

An ideal genealogical taxonomy would have a complete family tree of every organism and its offspring, but even such an ideal family tree is insufficient to determine which of these groups are species. Monophyly occurs at many levels of biological organization. Not all lineages of organisms are species. Ranking criteria are used to parse monophyletic groups into species.

Mishler and Brandon suggest choosing ranking criteria from the causal agent judged to be most important in producing and maintaining distinct lineages in the group in question.

> A species is the least inclusive taxon recognized in a classification, into which organisms are grouped because of evidence of monophyly (usually, but not restricted to, the presence of synapomorphies), that is ranked as a species

4.4 Species Membership and Genealogy

because it is the smallest "important" lineage deemed worthy of formal recognition where "important" refers to the action of those processes that are dominant in producing and maintaining lineages in a particular case. (Mishler and Brandon 1987, p. 406)

They believe that different lineages of organisms have different causal agents as ranking criteria, but think that there is one optimal ranking criterion for any particular lineage. Some believe that there may be more than one important ranking criterion for the same lineage, which could be divided into species in a different way (Kitcher 1984). Regardless of which interpretation of ranking criteria prevails, given the strict genealogical nature of the grouping criterion, the necessity of origin does not entail anything like a specific nature based on morphological characteristics typical to members of any given species. Furthermore, the judgment that a particular lineage is significant enough to be a species may depend on the number of organisms that might be grouped within that species or their relation to other organisms. The necessity of biological origins for a particular entity usually will not entail even that these other organisms exist, though their existence may play an important role in ranking the monophyletic group to which the organism belongs as a species. For these reasons, the species classification for a particular organism is not essential to it, or more cautiously, I know of no good argument that it is essential. The biological species of which a living individual of any kind is a part cannot play the role of a substantial sortal concept. Despite a long tradition in metaphysics, treating it this way just will not work.

The necessity of origin for biological entities guarantees that an entity is by necessity a member of whichever substantial kind it is actually a member. This alone does not entail that an individual will have the phenotype typical to the genetic makeup of its sex chromosomes or an essential nature derived from its species membership. This chapter concludes my treatment of identity at a time. In the next chapter, I will explore the persistence criteria for biological entities of the substantial kinds that I identified in Chapter 3.

5

Generation and Corruption

A substantial kind, as defined in Chapter 1, is a kind that an individual cannot cease to fulfill without ceasing to exist. In Chapter 3, I explained which kinds are substantial kinds for living things and the criteria of individuation associated with those kinds. Together, these chapters provide the means to establish the criteria of identity through time for living substantial individuals. A living thing of a particular kind begins to exist when an entity first exists that meets those criteria. That entity ceases to exist when it ceases to meet them. In this chapter I trace several representative life histories for living individuals of the substantial kinds I identified in Chapter 3 and elaborate on the relations between individuals of these kinds through time. In the following sections I explore the origins, growth, and eventual demise of each kind of living individual by considering a number of examples.

Toward the end of this chapter I examine whether death is necessarily the end of an entity. To this point I have treated death as if it were the irrevocable end of a living thing. I have assumed that the change from a living body to a corpse is one that an entity cannot survive. But I am convinced that this is not always the case – a living thing can die and then live again. If Walt Disney actually had been preserved cryogenically after his death and was later revived, he would still be Walt Disney.[1] If he was dead and is now alive, then our conception of the kinds of changes through which a living thing can persist needs to be revised. This didn't happen to Walt Disney but it happens every day to tardigrades, a microscopic aquatic animal. It is possible to preserve numerical identity through death under some circumstances. I explore these possibilities in section 5.4.

5.1 GENETIC INDIVIDUALS

A biological entity is a genetic individual if its parts share a common genotype because of descent without interruption from a common ancestor with that

5.1 Genetic Individuals

genotype. A new genetic individual begins when a cell develops a genotype distinct from that of its immediate ancestor or ancestors. Among higher animals, the sperm and egg are haploid and each possesses one of the numerous combinations of genes that the parent producing the gamete could produce. Each gamete is, itself, a unique genetic individual. A human being can produce 2^{23} (8,388,608) different combinations of chromosomes in a sperm or egg cell. This figure reflects only the normal process of reductive meiosis. Mutation, copying error, and crossing over introduce additional variations. The entity that results from the fusion of the gametes is a genetic individual distinct from either parent and either gamete.

When the gametes fuse, they are subsumed into a new genetic individual and cease to exist. That new genetic individual, initially composed of the zygote, persists from that time until no cell descended from that cell with that genotype exists. If there is a break in the continuity of genetic identity, aside from the replication during mitosis, that genetic individual cannot begin to exist again. During its existence the genetic individual is composed of the cells with that genotype that have descended from the first cell with that genotype, without regard for how those cells are arranged or the causal relations between them. As it gains or loses cells, however they are arranged, it grows or shrinks accordingly. Sometimes, as in the case of an asexually reproducing organism, what normally would be considered as reproduction should be treated as the growth of the genetic individual. I will discuss this view in greater detail below.

A genetic individual can grow in three possible forms. In the first form, the genetic individual is initially composed of a single cell. That cell splits into two cells by mitosis. If it is a single-celled organism, when it splits through mitosis the genetic individual is then composed of two single-celled organisms and their descendants as those cells themselves reproduce.

In the second possible form, the original cell divides by mitosis, and the resulting cells remain stuck together in a mass with no, or very limited, cellular differentiation and functional order. An example of this form of genetic individual is a mass of the bacteria that make the blue in blue cheese. The blue streak in the cheese is a bacterial colony. Although the cells tend to be massed together, there is no cellular differentiation. The cells are not causally integrated and removal of some of the cells does not have an adverse effect on the others. The growth of this mass of cells is the growth of a genetic individual. If some portion of the clump breaks off, the growth of that clump of cells and the original clump of cells is the growth of the genetic individual.

The third possibility is that the cells composing a genetic individual also compose one or more multicellular functional individuals. The normal course of metazoan development is just such a case. All the cells in a human body, excluding those from organ transplants or blood transfusions from anyone but your identical twin (if you have one) and mutations, are genetically identical descendants of the original cell formed by the union of your parents' gametes. All of these cells taken together compose a genetic individual. Any mutant cell lineages compose their own genetic individual.

Sometimes, the genetic individual is composed of multiple, multicellular units. Identical twins together compose a single genetic individual with spatiotemporally disconnected parts. The genetic individual can survive the death of either twin but not both of them. If one twin dies, the genetic individual would shrink to co-occupy the space occupied by the surviving twin. Each (detached) twin is a functional individual in his/her own right but together these two functional individuals compose a single genetic individual.

What we would normally call asexual reproduction is the growth of a scattered genetic individual composed of multiple functional units. Plants, fungi, and some animals can produce more than one functional unit that share a common genetic makeup through clonal growth, making them parts of a single genetic individual. Aphids reproduce asexually as well as sexually. All of the asexual generations of aphids together compose a spatiotemporally noncontiguous genetic individual. Janzen's image of the aphid life cycle can be used to make clear what composes a genetic individual among those species that reproduce asexually. I discuss Janzen's own argument during my analysis of the concept of an evolutionary individual, or unit of selection.

A genetic individual can also exist in some combination of the forms described above. A tribe of indigenous people in the Amazon basin are currently involved in a lawsuit to regain legal possession of cells taken from their bodies by a biological research company. These cells have been kept alive and reproducing on a petri dish. So there are genetic individuals composed of a rain forest inhabitant and some cells in a laboratory (Boyle 1996).

The genetic individual does not always look like what the layman thinks of as an individual. In some cases it will, as in the case of a genetically homogeneous human non-twin. In others it will not.

> The extraordinary study by Oinonen (1967) of the bracken fern *Pteridium aqualinium* has shown clones extending over 474 meters and suggests that the individual clones achieve ages of at least 1,400 years. Individual genets of such species must represent by far the biggest organisms (and individual biomass) of any living creature and must also achieve, in their long life cycle, the highest reproductive output of any known plant. (Harper 1985, p. 6)

Aspens, bamboo, bracken, and raspberries all regularly clone. The results are extraordinarily large and long-living genetically homogeneous entities. The genetic individual in cases such as these may be thousands of years old and weigh several tons.

5.2 Functional Individuals

Returning to the gigantic fungus from Chapter 2, the masses of *A. bulbosa* that were sampled are nearly genetically identical, and the best hypothesis is that this genetic homogeneity is the result of vegetative growth from a common ancestor. No tests were made to determine if there is a physiological connection between all of the genetically identical pieces identified or if the clone had become fragmented. The question of whether such physiological unity exists and is accompanied by the integration of parts rather than by mere connection must be resolved to determine if the clonal products form a functional individual as well as a genetic individual.

The genetic individual is not identical with any of the other units I have identified as natural kinds. If we think only about cases like a sexually produced non-twin, this is less than obvious. When we consider identical twins or a mass of clones we can see that the genetic individual differs in actual properties from any of the other units of individuation I have identified. Among higher animals (excluding identical twins), the genetic individual and the functional individual differ in modal properties. The genetic individual can survive the dissolution of the functional integration between the parts that compose it. The functional individual cannot. A functional individual's genetic makeup can (and does) change through time as mutations occur and genetic identity is lost. At any given time most multicellular functional individuals are composed of more than one genetic individual because of mutations and other sorts of copying anomalies. This makes the two units distinct even if their actual life histories do not differ at all. Because the units have different persistence conditions we cannot identify an entity under one substantial kind with an entity of a different substantial kind with different persistence conditions.

5.2 FUNCTIONAL INDIVIDUALS

A biological entity is a functional individual if the parts from which it is composed are integrated into a functional organic whole. A functional individual is composed of causally integrated heterogeneous parts. It would typically suffer impaired function if some of its parts were removed or damaged. It does not matter for our purposes whether it is the product of sexual or asexual reproduction, or even just stitched together on the spot. I am suggesting a general principle about functional individuals, according to which the current causal relation between the parts, not their history or genetic composition, matters.

The first functional individual I consider is a single-celled organism, for example an amoeba. The components of this cell are functionally integrated and form a functional individual.[2] The functional unit will almost always be spatiotemporally continuous, and there is a robust causal connection between its parts sufficient to causally integrate them. When that unicellular organism divides, the physically and causally isolated daughter cells are each a new functional individual.

A zygote is a functional individual. When the zygote begins to divide two things may happen. If the daughter cells separate, each begins a developmental process that results in distinct functional individuals, twins. But if the daughter cells remain stuck together, do they compose a functional individual continuous with the first one? This view tempts us because the zygote will eventually develop into a complicated and integrated metazoan body. My account would be simpler if the two cells formed a functional individual identical with the zygote, but there is good reason to resist this solution. After the initial series of cell divisions, the resulting cells are usually physically stuck together. Despite this connection, the mass of cells has no causal integration. Removing a cell does not have a significant effect on the other cells. Physical connection does not entail functional integration.

In the early stages of development, removing a cell has minimal effect on the other cells. Because of this, after each of the early cell divisions, the initial functional individual, the zygote, has been replaced by a small group of cells, each of which is a functional individual and none of which is identical with the zygote. The reason for this nonidentity is that when the zygote divides into two cells, the options are as follows.

i. It is identical with one or the other, but not both of them.
ii. It is identical with both of them.
iii. It has become a functional individual composed of two cells instead of one cell.
iv. It is not identical with either of the daughter cells.

Option (i) is an arbitrary choice between the two daughter cells, and identity is not based on arbitrary choice. It cannot be identical with both of them, (ii), because if that were true the transitivity and symmetry of identity would entail that the daughter cells are identical with one another, which is impossible because there are two of them. Because the pair of cells do not meet the criteria for being a functional unit, option (iii) is not true. They are not causally integrated.

The best option is (iv). The functional individual composed of the zygote is not identical with any later functional individual composed of aggregates of daughter cells or any single daughter cell, though it is d-continuous with them. Option (iv) raises a difficult question that is analogous to the question

5.2 Functional Individuals

raised about functional individuality in Chapter 3. How do multicellular functional individuals develop from parts (cells) that were not initially integrated? The simple answer is that the parts become sufficiently causally integrated. When, then, do they become causally integrated enough to compose a functional individual? Gradually increasing causal integration between the cells incorporates them into a multicellular entity. As with many sorites-like problems, it may be impossible to identify exactly when the transformation from group of autonomous cells to multicellular organism occurs.

That said, we can be more precise about when the multicellular functional individual begins to exist rather than remain content to say that it is some time after the first cell division and before the adult form is reached. The zygote, as we have noted, is a functional individual. When the zygote undergoes subsequent divisions, until there is enough causal interaction of the right sort to integrate that aggregate of cells into a multicellular functional individual, each daughter cell on its own is a functional individual. The normal metazoan life cycle then will involve a number of functional individuals before the cells are integrated enough in early development to compose a multicellular functional individual. For example, a fertilized sea urchin egg divides about once an hour in the earliest developmental stage (cleavage). These cells form into a hollow sphere called a blastula.

> The formation of the blastula is followed by a process known as *gastrulation* (from *gaster*, the Greek word for stomach), which gives rise to the primitive gut. Gastrulation in the sea urchin begins with the formation of the *blastopore*, an opening into the blastula. Cells near the blastopore break loose, and, with the aid of contractile pseudopodia with sticky tips, move over the interior surface of the blastula to the opposite pole. These cells are known as the primary mesenchyme. Next the entire cell layer closest to the blastopore turns inward, moving through the blastocoel to the opposite pole, forming a new cavity, the *archenteron*. The archenteron will ultimately develop into the digestive tract, and the blastopore will become the anus. (Curtis and Barnes 1989, pp. 923–924)

Somewhere during this process, the cells become integrated enough to form a functional individual.

> Once gastrulation is completed, evidence of cell differentiation can be observed. Cells in the mesoderm begin secreting calcium-containing granules that develop into tiny three-armed spicules. These become the supporting skeleton for the pluteus. At the point at which the endoderm touches the opposite side of the blastocoel, ectodermal cells curve inward to form the mouth of the larva. The archenteron subdivides into stomach, intestine, and anus, and a ring of long cilia forms near the mouth region. The single fertilized egg has become a number of specialized, differentiated cells, performing specific functions, such as

digestion; producing new organelles such as cilia; and secreting new products, such as the calcium containing skeleton. (Curtis and Barnes 1989, p. 924)

Vertebrate development follows a similar course during early ontogeny. Within five days of fertilization, the blastula is composed of about 120 cells in the shape of a fluid-filled ball. By the second week, the embryo has developed its body axis. In the third week the organs begin to form and the proto-heart starts to beat. By the end of the second month, the organs are almost complete and it is certain that a functional individual is in existence.

All living entities are characterized by a high level of causal integration at some level of functional organization. If there is insufficient organization between the cells to form a functional individual, the individual cell is integrated enough to be a functional individual. Among all living things, unicellular or multicellular, the cell is a functionally well-integrated unit. There are also well-integrated multicellular conglomerates.

Multicellularity appears to have independently evolved several times in evolutionary history and is the usual strategy for getting larger.

> If one looks at those organisms, primitive or advanced in structure, that live today, one must come to the conclusion that multicellularity is not something that arose once at some early time, but frequently, at many different times during the long course of the history of the earth. (Bonner 1988, p. 64)

How and why did multicellular life evolve? There are two common answers. The first is that getting larger has selective advantages and multicellularity is an easy way to get big.

> By becoming larger, the organism enters new size worlds where, among other things, they avoid predation and competition. On the other hand, any sustained selection towards size decrease would lead directly to size worlds of more intense competition, and therefore would be correspondingly rare. (Bonner 1988, p. 33)

Other advantages of multicellularity may include increased efficiency for some types of feeding, more effective dispersal, and protection from predators and external forces. There are things that an aggregate of cells can do that single cells cannot.

The second, more prosaic, explanation for multicellularity is that life started near the minimal rung of complexity and had nowhere to go but toward increasing complexity. If this is true, almost any random variation from this point would have to be toward increasing complexity and organization, even if there was no advantage to it and the shift was random rather than selection-driven. McShea (1996) provides a good review of this literature.

5.2 Functional Individuals

Whichever of these is the case, it makes sense that higher levels of complexity, instead of being produced from scratch, would be built up from simpler preexisting structures (Simon 1962). The best explanation of this struggle between lower and higher units resulting in the evolution of metazoan individuality is presented by Leo Buss in *The Evolution of Individuality* (1987). But see Jablonka and Lamb (1995) for a dissenting view.

According to Buss, the history of life is marked by the development of new self-replicating entities from the self-replicating entities that compose them. This hierarchical interpretation of evolution involving the interaction between entities at different levels of organization is central to Buss's view of individual development. He focuses on the evolutionary history of the relation between multicellular organisms and the cell lineages that compose them, but he recognizes that similar processes may have occurred during the evolution of the eukaryotic cell from symbiotic prokaryotes. A similar process is at least possible between unrelated multicellular organisms.

The evolution of multicellular functional individuality is a matter of degree. We know that well-integrated multicellular functional individuals evolved from single-celled organisms. Buss tries to explain how that transformation could have occurred through evolution by natural selection. Basically, his hypothesis is that competition and cooperation between cell lineages within colonies of clonal single-celled organisms led to the evolution of multicellular functional individuals, including the metazoan animals. He supports his hypothesis with evidence drawn from developmental biology and the variety of degrees of functional integration, cellular specialization, and germ-line sequestration found in contemporary organisms.

Variations in the lower unit (the cell lineage) may influence the replication rates of the lower unit and also that of a higher unit (organism) partially composed of the lower one. Some variations of the lower unit favor it at the expense of the higher unit. Cancer greatly increases the reproduction rate of the cancerous cells but usually has a negative effect on the higher functional unit. Some variations in the lower units – for example, a susceptibility to a replication-inhibiting enzyme produced by the organism – favor the higher unit at the expense of the lower. Some have a synergistic effect and increase the relative rate of reproduction at both levels.

> The organization of any unit will come to reflect those synergisms between selection at the higher and lower levels which permit the new unit to exploit new environments and those mechanisms which act to limit subsequent conflicts between the two units. (Buss 1987, p. viii)

Buss's hypothesis is that the history of evolution will be a reflection of these conflicts and synergisms. He presents strong evidence that this has been the case. He thinks that biologists have not previously recognized this evolutionary history because of the residue of Weismannian ideas about the early sequestering of germ-line cells. Though Weismann was wrong about the specific mechanism of inheritance, his view of the entire organism as a cohesive unit of selection continued to be accepted after his theory of inheritance had been replaced with modern genetic theory. Buss argues that this model of selection survived intact when it was combined with Mendelian genetics in the Modern Synthesis, despite the fact that it was a poor model for the development of the majority of living things, because the scientists who helped to create that new theory studied kinds of organisms that approximated the developmental characteristics Weismann had postulated.

Buss contrasts the development of the fruit fly with that of a hydroid, a small tentacled freshwater creature. During the first thirteen cell divisions of a fruit fly embryo, development is controlled entirely by directions from the maternal cytoplasm in the form of maternal RNA. After these divisions the developing embryo makes its own RNA to direct its own further development. By the time the fruit fly embryo's development is under the control of its own RNA, the sequestration of the cells destined to produce the gametes of the next generation has already occurred. To be heritable, a variation would have to arise during the first few cell divisions.

The development of a hydra is quite different. A hydroid zygote divides into an embryo composed of two populations of cells, the *I-cells* and somatic cells of various kinds. Somatic cells cannot become I-cells or a different kind of somatic cell. I-cells can give rise to any kind of somatic cell, undergo reductive division to produce gametes, or remain totipotent (able to generate a complete new organism). The totipotent lineage of a hydra may undergo an enormous number of divisions as compared with the Drosophila. Contrary to the model of inheritance characteristic of the Modern Synthesis, some variation occurring during hydroid development, that of any totipotent lineage, is heritable. If this is true, then the synthetic model of the individual as the sole unit of selection is wrong.

Nearly all of the multicellular protists and all plants and fungi develop by somatic embryogenesis. Nine animal phyla develop through somatic embryogenesis. The remaining animal phyla have epigenetic or preformative development.

> The ideal of the individual as an entity that may be treated as genetically uniform is at best an approximation. It is apparent that individuality is a derived character, approximated closely only in certain taxa. This fact is of substantial interest, for it means that not only is it inaccurate to consider the individual as the sole unit of inheritance in most taxa, but also that we have little assurance that it is

5.2 Functional Individuals

> appropriate to assume this to have been the case throughout geological time, even in those taxa in which individuality is now closely approximated. (Buss 1987, p. 20)

Because the scientists working on the Modern Synthesis studied animals with preformistic development, they neglected the other forms of development while formulating the Modern Synthesis and thus did not account for the variability that they ignored.

Buss believes that the Modern Synthesis has worked despite this blindspot because of what occurred in the evolutionary past.

> [E]volution has acted in all cellular-differentiating organisms in a manner such that individuality is approximated, albeit to varying degrees. Organisms appear as individual entities; evolution has yielded unambiguously discrete units. *The Modern Synthesis has stood as an enduring intellectual edifice for nearly a half-century because individuality has evolved.* (Buss 1987, p. 25)

The organisms studied in the formulation of the Modern Synthesis have preformistic development and sequestered germ cells because evolutionary pressures favored the properties that the biologists treated as primitives. The Modern Synthesis cannot explain the evolution of individuality characterized by genetic homogeneity and causal integration because it assumes individuality.

Buss proposes that at the origin of cell-differentiated multicellular life, the germ line was not determined from the beginning of development. There was competition between cell lineages within the same organism. Those lineages that prevailed at the cellular level had good or bad effects on the multicellular organism they partially composed, or they had no effect at all. If a lineage disadvantaged the individual, it and the individual harboring it usually perished. If it benefited the individual, it survived and was inherited. Later variants had effects not only on the whole they partially composed but on the cell lineages of earlier variants. Some later variant lineages suppressed or altered the effects of earlier ones. According to Buss, developmental processes evolved from interactions between these variants. "Those variants which had a synergistic effect and those variants which acted to limit subsequent conflicts are seen today as patterns in metazoan cleavage, gastrulation, mosaicism, and epigenesis" (Buss 1987, p. 29).

For the multicellular organism to evolve as a functional unit, some mechanism must prevent cell lineages from abandoning their somatic roles in the multicellular individual to increase their replication rate. Selection at the individual level must somehow control selection at the cell lineage level. How could this happen? Maternal cytoplasm controls early somatic determination. By the time the organism's own RNA affects development, the somatic lineages have been established and only the primordial germ cells remain undifferentiated to give rise to gametes.

Within the first 200 million years of eukaryotic cells, the basic metazoan body plans were already designed, and some cell lineages had lost the ability to produce a new individual. Cell replication no longer guaranteed heritability. The evolution of development can be seen as a record of past competition between different lineages. If it accidentally favors the multicellular unit it partially composes, it is selected for. The multicellular unit itself forms part of the environment in which competition between cell lineages can occur.

> Variants which further their own replication rate by restraining or directing the activities of neighboring cells may conceivably simultaneously enhance their own replication and the survivorship of the individual harboring them. Such variants, if they prove heritable, would produce an individual whose offspring would replay, as an epigenetic phenomenon, the same sequence of interactions between developing cell lineages. (Buss 1987, p. 78)

Metazoan development involves those lineages that favor both replication rates and somatic function. One method a cell lineage can use to further its relative reproductive success is *induction*. A lineage that can induce another lineage to adopt a somatic function while not doing so itself can thereby outreproduce that lineage. Similarly, a cell lineage that is less prone to induction than others will outreproduce those lineages that are susceptible to it if the somatic environment contains inducing cell lineages. The development of a contemporary organism consists of reciprocal inductions between cell lineages.

Buss's argument explains how selective forces could lead to the development of the highly integrated, complicated metazoan body. I include this lengthy summary of Buss's work for two reasons. His explanation provides a foothold sufficient for my metaphysical interest in the history of increasing functional individuality through the history of multicellular life as well as during the development of an individual organism. It also provides us with an algorithm to explain how once-distinct functional units can become a single functional unit.

There are other ways to form a functional individual. As the examples spread outward from the paradigmatic case of a genetically homogeneous multicellular organism, the distinctive properties of functional individuality will become clearer. There is a huge stock of examples from which to choose, and in the remainder of this section I will focus on a few of them.

In addition to genetic mutation of cells composing a multicellular organism, there are other ways that genetically homogeneous functional individuals can be produced. Some of these involve the aggregation of individual

5.2 Functional Individuals

cells. Some slime molds, for example *Dictyostelium discoideum*, originate as a group of individual cells that forage for food and divide into more free-living cells that do the same. When food becomes scarce something surprising happens.

> Certain amoebae become attraction centers, and the remainder of the population streams towards them. Soon, the random array is transfigured into rosettes of amoebae, with a rising center and radiating arms composed of amoebae still migrating inward. As the aggregation congeals further, it assumes a sausage shape averaging 1/2 to 2 millimeters in length. This new entity, called a pseudoplasmodium or grex, now performs like a multicellular organism. It has distinct front and hind ends, and moves slowly in the direction of light and heat. (E. O. Wilson 1975, p. 398)

The amoebae that form the grex are not necessarily related and may be genetically quite distinct. Nonetheless, they compose a functional individual that is distinct from any one of the amoebae, and in fact can behave in ways detrimental to the reproductive advantage of the individual amoeba during the formation of the spore-producing body.

Functionally integrated plants can also be composed of grafted parts that are genetically distinct, or even from different species. Apple and pear trees can be grafted together into a functional unit as tightly integrated as a tree grown from seed. Even more surprisingly, distinct genotypes can be integrated at the joint of the graft. Shoots grown from the joint of a graft between tomato, *Lycopersicum esculentum*, and black nightshade, *Solanum nigrum*, have resulted in bizarre plants with various combinations. For example, one cell layer of tomato covering a nightshade core and two cell layers of nightshade covering a tomato core were produced (Neilson-Jones 1969). Other types of chimera are also possible.

> Apparently in 1644, a Florentine gardener grafted a scion of sour orange onto a seedling stock of citron. The scion did not take successfully, but a bud arising out of the callus developed on the stock and grew up into a bizarre tree. On the same plant there were leaves, flowers, and fruit identical with the orange or with the citron, and there were compound fruit with the two kinds blended together or sectored in various ways. When propagated by cuttings, the tree retained its peculiar character. (Neilson-Jones 1969, p. 1)

On the assumption that a normal plant is a functional individual, there seems to be no reason to deny that chimeras such as these are also functional individuals. The parts of the plants do not share a common developmental history, but there is little to distinguish them functionally from a normal genetically homogeneous plant grown from seed. Xenotransplantation – the grafting of organs from one species of animal into another – also results in chimeral functional individuals. Pig and baboon organs have been transplanted into human beings with mixed success, resulting in a functional animal whose parts have an unusual history.

In addition to these kinds of functional individuals, there are some that are more controversial, for example cases in which a symbiotic relation develops to the extent that the diverse parts come together to form a functional individual. There are cases of both mutualism and parasitism that meet these criteria. The first case I describe is one that supports Lynn Margulis's theory that the eukaryotic cell evolved as a result of a series of symbioses when mitochondria, cilia, and photosynthetic plastids were absorbed by prokaryotes.[3] Kwang Jeon of the University of Tennessee was conducting research on a type of amoeba (the D strain of *Amoeba proteus*) when they became infected with a bacterium that rapidly reproduced itself within the amoeba's cell membrane (Jeon 1991). Most of the amoebae died. Those that survived developed interesting characteristics that separated them from the uninfected amoebae. They became more sensitive to starvation, temperature change, and overfeeding. They also became absolutely dependent on what were once their parasites for survival.

After just a few generations, the host amoebae depend on the symbionts and would die without them. The bacteria cause physiological changes in the amoebae and have been successfully incorporated as part of a functional unit.

The next case involves the more controversial claim that a symbiotic relation between two multicellular organisms can be intimate enough to justify calling the composite of the two a single functional individual. Some parasites alter their host's bodies in the same way that a cell lineage can induce another cell lineage to a purely somatic function, to the point that the host has been reduced to somatic function in the service of propagating the parasite. The host body has, in effect, been hijacked into service of the parasite.

Rhizocephalans are a group of parasitic barnacles. The larval rhizocephalan injects a cell or group of cells into a crustacean such as a crab, and consequently the single cell grows into a mass of rootlike fibers that absorb nutrients from the crab's body. Once the parasite has grown for a while inside the crab's body, it sticks its reproductive parts (externa) through the abdominal wall of the crab. The crab can reach the externa, but somehow the parasite alters the crab's behavior to prevent it from removing the externa. It then castrates its host, and, in at least one species, interferes with the crab's molting cycle so that it never molts again. This prevents the dislodging of the externa. The rhizocephalan then induces the crab to treat the externa as if they were the crab's own brood. The crab grooms the externa through gestation and then, when they are ready to disperse, the crab positions itself to release the

5.3 Developmental Individuals

parasite's larvae.[4] The female rhizocephalan has subsumed the crab's body to its own function.

To summarize these results, a new functional individual starts to exist when either:

1. Among single-celled entities, a free-living cell is produced.
2. A number of cells are caught up in a developmental process that causally integrates them.
3. Higher-level entities (organisms) are themselves combined in a way that causally integrates them, as in, for example, grafts or parasitism.

A functional individual ceases to exist when either:

1. It irrevocably dies.
2. The component of which it is composed becomes disintegrated.
3. That entity is subsumed into a higher-level entity or parasitized.

5.3 DEVELOPMENTAL INDIVIDUALS

A developmental individual begins with either a cell or small group of cells that will, unless interrupted, develop into a multicellular body. In the most basic case, a new developmental individual begins when a sexual or asexual propagule starts the genetically programmed developmental sequence resulting in a multicellular body. The developmental individual is composed of the single-cell propagule, its spatiotemporally contiguous daughter cells, and their descendants as they develop into a body.

The development of an adult metazoan body from a fertilized egg is the most straightforward example of a life history of a developmental individual. The basic idea is the creation of a new functional individual through a developmental process. This pattern can be repeated more than once within a life cycle. In Chapter 2 I discussed the asexual reproduction of aphids. All of the genetically identical cells composing those aphids also compose a single genetic individual. However, each layman's insect begins its life as a single cell, whether that cell is a sexual or asexual propagule. Each time there is a reduction to a single cell followed by a developmental process that results in a multicellular body, a new developmental individual is produced.

This concept can be expanded beyond metazoan development by the addition of multiple developmental processes within a single life cycle. The single developmental sequence of a metazoan animal may be multiplied several times over within more complicated life cycles that involve a series of radical morphological changes. The addition of more than one developmental sequence allows an organism to adapt to more than one environment or function – for example, underwater feeding and mid-air mating – without either developing a single body that does both things well or devising a way to change from one form to another while continuing as a viable multicellular organism. The reduction to a single cell or group of undifferentiated cells provides an opportunity to build an entirely distinct organism by initiating another developmental sequence.

In some species the young organism looks a lot like a small adult. As it ages it grows larger but other than that does not change very much. This kind of development is a rarity among insects.

> In almost 90 percent of insect species, however, a complete metamorphosis occurs, and the adults are drastically different than their immature forms. These immature feeding forms are all correctly referred to as larvae, although they are also commonly known as caterpillars, grubs, or maggots depending on the species. Following the larval period, the insect undergoing complete metamorphosis enters a pupal stage, in which extensive remodeling of the organism occurs. (Curtis and Barnes 1989, p. 576)

A caterpillar does not gradually change into a butterfly. The caterpillar's organs are consumed by the new butterfly body to fuel its development. The butterfly body begins to develop when hormones trigger the growth of imaginal discs, relatively undifferentiated groups of cells that then grow into the parts of the distinctively adult body. The larva and the adult insect are distinct developmental individuals. In the metamorphosis of a caterpillar into a butterfly, the earlier developmental individual is destroyed in the creation of a new one, but this is not necessarily the case for other species. An aphid is not destroyed in producing an asexual propagule. Both can coexist simultaneously and continue independent developmental processes.

Even more dramatic transformations can occur. A parasite that tracks different hosts at different stages in its life cycle presents an odd case. The rhizocephalan, discussed in the section above, is such an entity. The female rhizocephalan begins its life as a nauplius, a free-living larva. The nauplius turns into a cyprid larva that attaches itself to a crustacean host and metamorphoses into a kentrogon.

> The kentrogon is smaller and simpler than the cyprid, but develops one crucial and special organ – Delage's "dart" (now generally called an injection stylet). The kentrogon's dart functions as a hypodermic needle to inject the precursors of the adult stage into the body of the host! (Gould 1996, p. 14)

In some species of rhizocephalan only a single cell is injected into the host. Gould says of this remarkable bottleneck that it is "as though, within the rhizocephalan life cycle, nature has inserted a stage analogous to the fertilized egg that establishes minimal connection between generations in ordinary sexual organisms" (Gould 1996, p. 15). The injection marks the beginning of a new developmental individual, which spreads through the parasitized crab's body.

A developmental individual begins its existence as a single cell or a group of undifferentiated cells. These cells develop into a multicellular body. The developmental individual persists as long as the developmental process is occurring or the mature body resulting from that developmental process continues to exist.

5.4 RAISING THE DEAD

The continuation of life is insufficient for continued numerical identity. Nothing dies when a single-celled organism divides into two new daughter cells, but despite the continuity of life, the initial organism ceases to exist after the division. In this section, I address whether the continuation of life is a necessary condition for persistence and suggest that under special circumstances a living entity can persist through the loss of its life. The types of cases that I have in mind are ones in which a biological entity has been put into stasis or otherwise preserved and then later revived. It seems appropriate to describe a case like this as one in which an entity was alive, was not alive, and then was alive again while remaining numerically the same thing.

My concern for this topic is primarily derived from the forms of stasis that preserve the structures of a living organism intact or in restorable form, but prevent those structures from functioning. It is currently possible to freeze and then later revive sperm, eggs, and even zygotes. Some people have had their corpses frozen in hopes that they can one day be revived. Freezing and equivalent forms of stasis are particularly attractive candidates for preserving identity without a continued life. Freezing makes the body a nonmetabolic object. Perhaps the best example of what I have in mind though is tardigrade cryptobiosis. Tardigrades are microscopic aquatic animals.[5] In addition to being able to survive freezing, some species of tardigrades, such as *Richtersius coronifer*, can survive complete dehydration or anhydrobiosis.

> Anhydrobionts achieve a degree of dehydration that not only involves the loss of 'free water,' which forms the aqueous solutions of the body, but also the loss of 'bound water,' which is required to maintain the structure of vital hydrated macromolecules such as proteins, membrane phospholipids and nucleic acids. (Kinchin 1994, p. 79)

Though not completely established, it seems most likely that during this state there is no metabolic activity in the tardigrade, yet it can be rehydrated and resume living.

Peter van Inwagen (1990) has written about the possibility of a life ceasing and then starting again. Van Inwagen's basic thesis is that the only composite objects in the world are biological individuals. If we translate parts of his work to a framework that is compatible with this work, we can fairly interpret him to be referring to functional individuals. He believes that an entity exists only so long as the activities of the simples (noncomposite parts) that compose it constitute a life. "If an organism exists at a certain moment, then it exists whenever and wherever – and only when and only where – the event that is its life at that moment is occurring" (van Inwagen 1990, p. 145). This view naturally leads to the question of whether a life can stop and then start again while still being the same life. If this is possible, then on van Inwagen's view, the revived entity is numerically identical with the entity that ceased to live before. He gives us a specific example to consider.

> Suppose we take a healthy cat and freeze it; suppose we reduce its body temperature to very nearly absolute zero by some technique (not currently available, by the way) that does it no irreversible organic damage. Suppose we then revive the cat. It seems clear that the revived cat is the cat we started with.... But it also seems clear that the cat's life ceased when it was frozen. Life is the sum of a great many chemical processes, and no chemical processes at all are going on inside the frozen cat. And, therefore, one might argue, the friends of *Life* must accept the possibility of discontinuous lives. (van Inwagen 1990, p. 146)

I think that the frozen cat is clearly not alive, though it once was, which makes it dead. My view becomes more convincing if we do not revive the cat and leave it frozen forever. The state of being dead seems to me to depend on the current relations of the parts of an organism rather than on what happens to that organism in the future.

Van Inwagen is not as sure as I am that the cat's life ceases when it is frozen. He does not think that the cat is definitely dead, but he leaves that judgment to the reader. Instead of pushing the claim that the cat is still alive, he argues that there are two ways that a living entity can cease to live. It can be either suspended or disrupted. A life is *suspended* if "it has ceased and the simples that were caught up in it at the moment it ceased retain – owing to the mere absence of disruptive forces – their individual properties and their relations to one another" (van Inwagen 1990, p. 147). A life that has ceased in any other manner is *disrupted*. According to van Inwagen, a disrupted life can never start again as numerically the same life. He alters his answer to the special composition question to incorporate this addition.

5.4 Raising the Dead

(∃y the xs compose y) if and only if

the activity of the xs constitutes a life or the individual properties of the xs and their relations to one another are unchanging (at the level of activity at which the processes of life take place) and when the xs were last changing, their activity constituted a life. (van Inwagen 1990, p. 148)

The distinction between a suspended life and a disrupted one is useful, but there are better ways to draw it. Unless being suspended entails that the entity composed of those simples can be revived, van Inwagen needs to say more about the arrangement of the simples, such as adding a clause about the life being restorable, which is clearly the distinction he intends to draw. Without that proviso, he has not yet provided us with sufficient conditions for continued identity of a life.

Nor do I think that he has provided a necessary condition for an entity to survive its (temporary) death. It is not plausible that just any change in the relations of a frozen entity's parts to one another would disrupt the life of the entity so that it could not be revived. For example, the removal or replacement of a single atom is not an identity-destroying change. Perhaps some form of freezing will prove to be useful in organ transplant surgery. An entity is completely frozen. Its damaged and now frozen liver is removed, and a healthy frozen liver is implanted. Once the cryosurgery is completed, the resulting entity is thawed out and restored to life. I suppose that I am prejudicing the case by saying that it is "restored" to life, but isn't that really what has happened?

Would it destroy the identity between the life of the cat before it was frozen and the life of the cat that is thawed out, if, while the cat was frozen, we sawed it neatly in half and then stuck it back together into exactly the same way we found it? No. Nor would (carefully) cutting the frozen cat into a dozen pieces and then sticking them back together and then thawing it out. Both of these "operations" seem survivable. So, it seems as if it is not actually essential that the parts remain in exactly the same orientation. The issue is one of preserving identity. If the conditions that van Inwagen chose are not correct, which ones are? Such conditions must preserve numerical identity between the creature that was frozen and the entity that is thawed and revived.

A mechanical analogy may be helpful here. Most machines can continue to exist whether they are in operation or not. In the *Essay*, Locke compares a living animal to a watch.

> Something we have like this in Machines, and may serve to illustrate it. For Example, what is a Watch? 'Tis Plain 'tis nothing but a fit Organization, or Construction of Parts, to a certain end, which, when a sufficient force is added to it, it is capable to attain. (*Essay*, Book II.xxvii.5)

Sometimes the watch functions and sometimes it does not, but it remains numerically the same machine either way. The crucial difference between such a machine and a living entity is that, under normal conditions, most machines can be started, stopped, and then restarted. The same cannot be said for a living entity. When a living entity stops functioning, the "machinery" of its functioning is usually irreparably damaged. But this is not always the case. Some means of stopping some lives leave them in a condition much like an unwound watch.

Freezing can preserve the mechanisms of life while ceasing their operation. If we accept that an entity retains numerical identity through a perfect and reversible process of stasis, there are some other possibilities to consider. If an entity is in a condition of perfect stasis, then any reversible process to which we subject the perfectly preserved entity should also not interfere with its continuing identity. If it is carefully cut in half or quarters for storage and could later be put back together, it is preserved. Also, small modifications, additions, or subtractions of atoms or even some organs do not seem to interfere with its persistence.

What if the entity is not actually revived though it could have been? If it could have been revived, then it has retained the necessary structure and is like a machine that has been turned off or disassembled. So long as we *could* turn it back on, I think that it continues to exist. What does 'revivable' mean? It may be currently technologically impossible to revive some entity that could be revived by a soon-to-be-developed method. The ontological status of the suspended entity should not be dependent on the level of technology available at the moment. Instead, whether or not numerical identity is retained should depend on the possibility or impossibility of restoring the body to life, whether it actually happens or not. This position may open a gap between what anyone thinks is possible and what is possible, but this causes no special problems for my view unless we are presumed currently to know everything.

If my conclusions in this section are true, theories of identity that treat having the same life as a necessary condition for being the same entity cannot work, because entities can, under some conditions, persist after death. The exact boundaries of revivability are not currently known, but revivability clearly extends further than a theory based on continuing life can comfortably explain.

6

Personal Identity Naturalized

Our Bodies, Our Selves

To this point, I have avoided the topic of personal identity in my analysis of biological individuality. I did this for two reasons. Focusing on personal identity might have deformed my account of biological individuality by encouraging me to develop a model into which personal identity could easily be inserted. The second reason is that I have never been sympathetic to philosophical theories of personal identity that treat a person as a substantial individual distinct from a biological organism (functional individual). I think that the important issues of human identity can be resolved without recourse to treating *person* as a substantial kind. Now that I have developed a system of individuation though, it seems appropriate to explore its application to human beings and personal identity. I intend for my account of biological individuality to provide the grounds for individuating all living things, so it should apply to human beings. In the sections below, I discuss the different kinds of biological individuality as they apply to human beings. This discussion provides the groundwork for an exploration of the connection between personal identity and the biological identity of a human organism.

6.1 HUMAN BEINGS AS BIOLOGICAL ENTITIES

In addition to explaining the nature of biological individuality for any number of other organisms, a system of biological individuation must provide an account of human individuation. A human being is a higher animal. As a higher animal, a human being is one of the easier living things to individuate and trace through time. In this section I explain the different kinds of biological individuality as they apply to human beings and describe the persistence conditions for each. In the next section, I explain how these kinds relate to personal identity.

A new human genetic individual begins with the fusion of a sperm and an egg. The resulting zygote combines aspects of the genetic structure of its mother and father and is genetically distinct from either of them. That human genetic individual persists for as long as cells with that genotype directly descended from a cell of that genotype continue to exist. What usually happens is that the zygote divides into genetically identical daughter cells. The daughter cells remain stuck together as they continue to divide. The result is a mass of genetically homogeneous cells. Simultaneously, the cells are caught up in developmental processes that typically result in the production of an integrated multicellular body. The genetic individual is composed of the genetically identical cells that compose that multicellular body. Because of mutations and other copying errors, the multicellular body may not be genetically homogeneous. It may be composed of cells that also compose more than one genetic individual.

Monozygotic twins present a slightly more complicated case. Cells from the first few cell divisions after the zygote is formed become physically and causally separated from one another and continue to divide. Each group of cells then begins a distinct developmental process resulting in a multicellular body. Both bodies originate from the same zygote, so, barring mutations, the cells that compose these distinct bodies are genetically identical and also compose a spatially scattered genetic individual.

A human zygote is also the first stage of a new developmental individual. That developmental individual persists for as long as the developmental process continues or the adult multicellular body exists. In the earliest stages of development the group of cells that compose the developmental individual are stuck together, but they are not causally integrated. Nor does removing one cell have an effect on the developing organism, because of the lack of functional integration in early development. Eventually, those cells differentiate and develop into a functionally integrated multicellular body. Although there may not be a precise moment at which it occurs, at some time during this developmental stage the cells that compose the developmental individual become functionally integrated enough to also compose a functional individual.[1]

A human functional individual persists until one of several things happens to it. A violent death may cause the physical and causal separation of enough of the body's parts that the functional individual cannot maintain the processes involved in the continuation of life. Alternatively, a vital organ may be destroyed or impaired by disease, causing the ultimate breakdown and death of that living human body. As with the time of origin, it will often be impossible to specify an exact time of death for the multicellular functional individual without making an arbitrary decision.

6.1 Human Beings as Biological Entities

Though I know of no such cases, there are two other ways in which a human functional individual would cease to exist. The body could lose its functional integration, though its parts continued to live. The causal interactions between the cells that compose it could cease while the individual cells themselves were kept alive and reproducing.

A functional individual would also cease to exist if it were caught up in a higher-level functional individual, like the cell lineages that Buss discusses. It could also be parasitized to such an extent that it was completely subjugated as a part in the service of some other entity. Some might argue that some human social structures, for example a football team or a fascist state, should count as cases of higher-level functional individuals that subsume the multicellular functional individual and so destroy its individuality, but this seems to be overstating the degree to which such social structures actually dictate our behavior. A player is not integrated into a team in the same way a cell or an organ is integrated into a body. I know of no parasite of humans that could manipulate us to the extent that a root-head manipulates a crab, but the thought of it happening is so fascinatingly unpleasant that it has been the subject of horror movies and at least one episode of *Star Trek*.

A vague boundary divides the losses or changes in parts that a functional human individual or other higher animal can survive from those that one cannot, but it may be useful to approach this boundary from both sides even if we find that they are separated by a no-man's-land rather than a crisp border. A human functional individual can certainly survive at least some changes in the matter of which it is composed. Normal biological processes involve ingesting, digesting, and metabolizing food and excreting bodily wastes. A functional individual can gain matter by a variety of processes such as normal growth, getting fat, or body building; all of which increase the amount of matter that compose that functional individual. The body can also shrink as muscles atrophy or fat is consumed by metabolic processes. This kind of change, the addition, loss, or exchange of small parts, is always occurring if we are alive.

Other compositional changes may be more suspect. These are cases in which larger, more organized parts such as digits, limbs, or organs are lost or replaced. It seems obvious that a human functional individual can survive some loss of parts. Losing a toe to the lawnmower or an ear in a barroom brawl are regrettable losses, but losses that the functional individual survives. Nose jobs, liposuction, loss of a kidney, lung, or appendix are all survivable too. This intuition is supported by the fact that the body continues to function as a whole despite these changes. Some of these parts do not matter much, and losing them may cause the organism no problems. The loss of fifty pounds of fat may actually help the organism to function better. Other losses, such as that of a digit or a lung, do impair the function of the organism of which it was a part, but it is built into the definition of functional individuality that the removal of parts typically causes impaired function. The fact that an entity's function has been impaired does not mean that the entity does not survive the loss.

Just as a human functional individual can survive the loss of some of its parts, it can also survive the addition of parts if those parts become incorporated into it. If a human being's finger has been severed, it is no longer a part of the functional individual because it is not functionally integrated with the other parts of the organism. If it is appropriately reattached, it once again is a part because it has resumed the sort of integration into the organism that it had before being detached. The fact that the finger was previously a part of the functional individual plays no important role in this story. What matters is the current relations between the parts. Receiving an organ transplant does not interfere with the continued identity of the transplant recipient. The transplanted organ assumes a functional role in that organism and is thereby a part of it. By extension, multiple organ transplants do not seem to interfere with the persistence of the functional individual either, though at some point we may face Ship of Theseus–type problems, but these are hardly unique.

I am less certain about what to say about the introduction of artificial parts such as a pacemaker or artificial heart into a functional individual. The part is nonliving and made of artificial components. A natural organ is itself alive and, at least at the cellular level, it is composed of living parts. Below the cellular level of organization though, both the artificial and natural organ are composed of nonliving parts. When implanted successfully both an artificial heart and a natural heart become functionally integrated into the living body, just as the organism's original heart was.

Not all parts can be changed without the loss of identity. I do not think that a human being or any other higher animal would continue as the same functional individual after a brain transplant. This is a common intuition and it can be justified by an argument. Brain transplants are currently technologically impossible. Because of this impossibility, when we think brain transplants provide us with evidence, we should draw philosophical conclusions with caution. Some of what we believe would happen in such a case may be the product of our prejudice. It would be wise to examine what those presumptions are. Foremost among them is the belief that if we revived someone after a brain transplant, he would profess to be the brain donor rather than the brain recipient. He would claim to remember things that had happened to the human whose brain it previously was. We are less sure about what would happen to the athletic ability, coordination, or disposition of the functional individual with the brain transplant because it seems as if these attributes would be

6.1 Human Beings as Biological Entities

influenced by parts of the body other than the brain, though the brain plays a role in them. I think that it is worth pointing out that these speculations may be entirely wrong, but we can work with them as tentative assumptions.

If it works at all, though, we can be sure that the new brain would regulate and direct bodily functions that were previously controlled by the replaced brain. It would coordinate the activities of the other parts of the body. The brain is the central hub where information comes in from the nervous system, is processed, and responses are sent out. For these reasons, the brain seems to be a special organ with properties that distinguish it from any other organ of the body.[2] If you lose your brain you are truly lost.

A further argument could be made for accepting the maxim that you go where your brain goes, though this does not follow from the fact that losing your brain means losing yourself. One line of approach is to argue that under some sets of circumstances a human being could be composed of just a brain. Peter van Inwagen presents one way to think about brain transplants that makes this view plausible. Starting with the idea that a human being could survive the loss of a digit, he explores just how much of that human could be carved away without it ceasing to exist. He describes a brain being kept alive in a vat.

> It, the living object in the vat, was recently a 150-pound, four limbed object. It has shrunk, or, rather, has been whittled down. It was once a normal man. It is now a radically maimed man, a man who is about as radically maimed as it is possible for a man to be. The case of the brain in the vat is logically not much different from the case of a man who has lost an arm: The latter was recently a 150-pound man and has lost about six pounds of bone and blood and tissue; the former was recently a 150-pound man and has lost about 147 pounds of bone and blood and tissue. (van Inwagen 1990, pp. 172–173)

This story captures an important distinction between the brain and any other organ. If we separate a brain from its original body and keep both alive, it is the body, rather than the brain, that is in a position analogous to that of a detached finger. The brain entirely composes the human being in this situation, though, as van Inwagen notes, that human is radically impaired. The operation that we described as a brain transplant may more accurately be described as a body transplant rather than a brain transplant. Just as the detached finger or transplanted organ is incorporated into a functional individual, the body would be incorporated as parts into the functional individual (human being) that was previously composed of just a brain.

Unlike transplanting a brain, which we will assume is only technologically impossible, transplanting either a single memory or an entire set of memories is a conceptual impossibility. It is not conceptually impossible to transplant an apparent memory or set of apparent memories, but appearing to be a memory does not entail, even in everyday cases, that something is a memory. I think that it would be worthwhile to say something about memory, because memory has misled people into odd views of personal identity. Memories are true. A false memory is, strictly speaking, a contradiction. If you remember an event, then that event actually happened. If you remember something happening, then you existed when the event occurred. Further, for something to be a memory, there must be the right sort of causal relation between the actual event and your apparent memory of it. Theories of personal identity that make memories or apparent continuation of consciousness detachable from bodily identity are mistaken. The mistake lies in conflating an actual memory, which has the right sort of causal connection to the actual event and the body of the person, with an apparent memory, which lacks that connection. This is, of course, a stipulative sharpening of our notions of what constitutes a memory. Much of what I argue for here could be reconstructed in a position that allows for false memories, but I have seen no convincing position that allows for false memories which also deals effectively with the problems posed by duplication when memory is not tied to a particular organism. An apparent memory may be internally indistinguishable from a real memory, but can fail to be a memory in two ways. The event that seems to be a memory may not have occurred. Alternatively, there may not be the right sort of causal connection between the human with the apparent memory, the apparent memory, and the event.

If a human's memories were transplanted into someone else's brain as apparent memories, the memory transplant recipient is not thereby identical with the human whose memories were transplanted. This fact is easy to see if the "memory donor" has her memories copied but not destroyed and survives the process with memories intact. Whether the donor survives with memories intact, has apparent memories implanted, or is physically destroyed in the process does not affect whether or not a real memory transplant has occurred. Regardless of what happens to the donor, a transplant of real memories is impossible. If we could transplant simulacra of one human's entire memory into someone else, why could we not transplant those memories into more than one person? Each of these people may claim to be the original person and feel just the same way about the claim, but none of them would be identical with the original person. Similarly, if we could transplant a single apparent memory from one person – for example, blowing out the candles on the cake for one's thirtieth birthday – each person into whom we transplanted this memory would seem to remember blowing out those candles himself, but that does not make it so. Because there is no necessary connection between apparent memories and the persistence of a human being, there is no reason to believe that anything survives of a human if all of his real memories are transferred as apparent memories before his death into another human being whose memories have been erased. The result of this experiment would be a dead human being and a human being deluded into thinking he is the dead man. Friends of the dead man might prefer the company of the simulacrum to that of a stranger, but this is hardly evidence that it is that person.

Though it is technologically impossible, I do not know that it is a conceptual impossibility to transfer apparent memories from one person to another. For the sake of argument I am willing to grant that it is possible. Even if there were no problems in resolving cases of personal identity in which sets of apparent memories were duplicated and transplanted into more than one person, I do not find theories of personal identity based on apparent memories persuasive. Someone with an apparent memory may think that the content of the apparent memory happened to him, but this does not make it so, whether the apparent memory involves an isolated event or an apparent life full of events. An apparent memory is not sufficient for continued identity in the way an actual memory is because it is dissociated from the context that makes an actual memory a memory, and part of that context is the continuation of bodily identity.

None of the higher brain functions such as consciousness or memory are essential for continued identity of a human functional individual. Some human functional individuals may never be conscious, though they are still functional individuals. If a human being goes into a coma it does not cease to exist. A human being who as a result of either congenital defect, accident, or disease loses some or all higher brain function, while retaining those parts of the brain that control the metabolism and involuntary bodily functions, continues to exist. In the next section I want to explore what reason there may be for introducing a further substantial kind to this account of human individuation.

6.2 IS A PERSON A HUMAN BEING?

I think that a human being is a human functional individual. This fits in well with our naming practices for humans. In my discussion of what it is to be a human functional individual I did not explore many properties typical of a human being. Have I neglected properties significant enough to justify the addition of another substantial kind to the ontology of living things? As a first step toward answering this question I will explain some properties of the human functional individual.

If a human being develops in the typical fashion she gradually acquires a number of interesting properties between birth and adulthood. Among these properties is consciousness, which may be better described as a complex set of properties that allow her to have subjective awareness about her surroundings, and self-consciousness, an awareness of her own consciousness. A human being gains the capacity to know something about its situation in the world and can report on some, but not all, aspects of its progress through the world.

There are definite limits to a human being's awareness and self-awareness. It has no firsthand knowledge about what happened to it after it began to exist but before it was born. It has no direct knowledge of what happens to it while it is unconscious or asleep. Some of its own abilities and cogitations may remain opaque to its self-awareness. A human being may be unable to explain the precise series of muscle contractions involved in a swan dive, though she can do a perfect swan dive. She may know that the sound she heard came from behind her and to the left, but still be unable to describe how she knows this.

Among the conscious abilities of a typical human is the ability to remember things that happened to her. She develops something like a coherent narrative of her own life so far. She accurately remembers a large number of episodes from her past, but does not remember many of the other things that did happen to her. Among these are things that happened while she was young, under anesthesia, asleep, knocked out, or experiencing a mental or physical trauma. Human beings can form mental states that are internally indistinguishable from memories but are not memories. Sometimes they are not memories because the events that they putatively record as having happened to the human did not happen. Cases like this can range from everyday slipups to thoroughly fictitious accounts of important or traumatic events. Alternatively, the events in such an apparent memory may have actually happened to that human but are now known by her to have happened by reference to diaries or an old yearbook rather than by direct recollection.

A human being usually presents itself as a unified being. We tend to act in a manner consistent with our past actions. We typically make plans for the future and even carry out some of those plans. To a significant extent the human being functions as a single unit of behavior. It can move in a coordinated fashion with all of its limbs apparently under common control. There are many exceptions to this model of unity. I will only touch on some of them here. Humans are subject to akratic states – for example, the simultaneous desire to smoke a cigarette and not to smoke a cigarette. Alterations in the normal structure of the brain can result in fascinating breakdowns of this normal unity. Under quite limited and temporary experimental conditions, a human whose corpus callosum has been severed can behave in ways that suggest that some parts of his body are under the control of one brain hemisphere and others are under the control of the other hemisphere.[3]

6.2 Is a Person a Human Being?

Other typical properties of a human being are developed through relations with other human beings. A normal human being will develop a moral character as part of her social interactions. Without the right sort of relations with other humans, a human's social development may be impaired. Part of her self-conception involves her relations with parents, siblings, friends, lovers, and co-workers. Her normal life involves interacting with other human beings in a complex web of social relations. The social relations between human beings and the extent to which some of the human potentialities have been realized will to some extent determine how other humans think that they should treat that human being and how they actually do treat her.

In this brief discussion of some of the characteristics of a typical human being I was careful not to make reference to the term 'person,' and I tried to avoid the suggestion that there was something called personal identity that should be treated as distinct from the identity of a human functional individual. Many people use the terms 'person' and 'human being' interchangeably and mean exactly the same thing by both terms. Others think that there is an important distinction between the two. I do not think that this dispute can be resolved by looking at the dictionary. The important issue is to determine if there is some further substantial kind, whatever term we use to identify it, that picks out substantial individuals distinct from human beings. That is to say, does something really cease to exist when it loses some of the properties that we associate with a fully functional human being?

When I discussed the properties of a normal human being, I was careful to note that they were typical features. No human being has those properties throughout its life. Some human beings never have them at all or lose them well before they die. Losing these properties or failing to attain them is something of a tragedy because it represents a significant failure to achieve potentialities that make for a fulfilling human life. Losing the properties of consciousness or self-consciousness is perhaps a tragic loss, and one that will clearly affect how that human functional individual will be treated and probably ought to be treated by other human beings, but it is still a human being, it has not ceased to exist. Its animal life continues in a diminished capacity, but it continues nonetheless.

The normal or abnormal course of human life can be described without mentioning persons as distinct from human beings. A human being can lead part of its life without the properties that I described, acquire them, and then lose them again in senility without ceasing to be a human being. Presumably, the main contenders for another substantial kind would deal with the properties of consciousness, memory, or the moral status of an agent. As such, someone who supported these characteristics as marking a kind of substantial individual that cannot survive their loss would argue that individuals of that kind were not identical with human beings. In the remainder of this chapter I will consider some of these views and explain where I think they go wrong and how what they get right can be incorporated into my view. I am sure that I will not convince a die-hard adherent of an alternative position, but I can make a strong case for my view.

Some philosophers, Descartes foremost among them, would argue that my account of human identity leaves out something important, the identity of the person. Personal identity in this sense of the word is distinct from the identity of a living human being because it can persist while detached from any material body.

The literature on this nonphysical conception of personal identity is too extensive for me to discuss here, but I want to explain the basic mistake involved in this view. Descartes uses a thought experiment to justify his belief in a nonmaterial mind. He starts by asking how much of what he sees around him could be illusion. His conclusion is that the existence of the entire material world, including his body, could be a deception.

> How about eating or walking? These are surely nothing but illusions, because I do not have a body. How about sensing? Again, this also does not happen without a body, and I judge that I really did not sense those many things I seemed to have sensed in my dreams. How about thinking? Here I discover that thought is an attribute that really does belong to me. This alone cannot be detached from me. I am; I exist; this is certain. But for how long? For as long as I think. Because perhaps it could also come to pass that if I should cease from all thinking I would then utterly cease to exist. . . . I am not that connection of members which is called a human body. (Descartes 1984, *Second Meditation*)

His argument is based on the fact that he could imagine that he could continue to exist as a nonmaterial thinking thing even if the material world did not exist. I discussed Descartes's argument for this conclusion in my criticism of thought experiments in Chapter 1. Because he has the ability to imagine this as a possibility he thinks that it must be a real possibility. Imagining that something could be the case is not adequate evidence that it really could be the way that it is imagined to be. This is particularly true with cases like this where the underlying relation between the things that we are speculating about is unknown. Some of the things that we think we can imagine are impossible. We think that they are possible because we are ignorant of some important fact about the situation. I discussed several examples of this kind of reasoning in Chapter 1. For example, if we do not know some important facts about the relation between them, we might imagine that Hesperus could cease to exist while Phosphorus continues on as usual. Similarly, though I feel that I could imagine the scenario that Descartes describes, if consciousness

6.2 Is a Person a Human Being?

has a material basis in the brain, and we have every reason to think that it does, our consciousness could not exist if the material world, including our brains, was eliminated.

Descartes also conflates thinking about ourselves and thinking about our conscious states. Conscious states are real and important, particularly to whomever's conscious states they are. It is certainly true of anyone who is reading these pages that she is a thinking thing. Thinking is among the things that typical human beings can do, but I see no reason to identify the thinking being as one that ceases to exist when it ceases to think. To do so seems to me to confuse an attribute of a thing with the thing itself. I can think about myself and I can think about other things as well. When I think about myself, I may think about other aspects of my mental life, my desires, or my imagination, but I may also think that I wish that I could slam-dunk a basketball. It seems bizarre to me to think that in these cases I am talking about something distinct from myself. I am talking about me. I see no compelling reason to adopt the view that persons are immaterial thinking things commonly associated with human beings.

Nor is there good reason to think that the limits of our thinking mark the limits of our existence. I am a living human being. I currently have the ability to think about other things and to think self-consciously about myself. I existed without that ability, and I may lose it before I lose my life. In that case, I am no longer able to observe my own life, though it continues. I may regain that ability and so regain awareness of my life, but I do not in virtue of regaining this ability start a new life.

Locke's view on personal identity suffers from the same problem of isolating a set of properties of a human being from the human being and then conceiving of those properties as necessary and sufficient conditions for personal identity without regard for the persistence of the human being. Locke presents a mechanist view of the persistence of plants, animals, and human animals but departs from this view when he examines personal identity.

> [We] must consider what Person stands for; which, I think, is a thinking intelligent Being, that has reason and reflection, and can consider it self as it self, the same thinking thing in different times and places; which it does only by that consciousness, which is inseparable from thinking, and as it seems to me essential to it. (*Essay*, II.xxvii.9)

He clearly believes that there is a distinction between personal identity and the identity of a living body.

One reason that Locke is concerned to distinguish a person from a human being is the connection that he sees between having the same consciousness and being worthy of moral praise or blame. "In this personal identity is founded all the Right and Justice of Reward and Punishment; Happiness and Misery, being that which everyone is concerned for himself" (*Essay*, II.xxvii.18). Locke recognizes that a human being may reasonably not be held accountable for all of his actions under some circumstances. Locke describes an imaginary scenario to illustrate this point.

> If the same *Socrates* waking and sleeping do not partake of the same *consciousness*, *Socrates* waking and sleeping is not the same Person. And to punish *Socrates* waking, for what sleeping *Socrates* thought, and the waking *Socrates* was never conscious of, would be no more of Right than to punish one Twin for what his Brother-Twin did, whereof he knew nothing, because their outsides were so like, that they could not be distinguished; for such Twins have been seen. (*Essay*, II.xxvii.19)

We do not need to turn to such a fantastic case to see his point. Some realistic examples of the same sort might include whether or not to consider someone morally responsible for what they have done in a fugue state, while sleepwalking, or under the influence of a drug. Because we wish to punish or praise humans for what they intentionally do, there is good reason to separate some of our actions from others on this basis. When we examine someone's actions to determine to what degree he should be held morally responsible for them, we do more than determine if those actions were the actions of his body. We want to know about his mental state.

These good forensic reasons for adopting Locke's view on personal identity can easily be incorporated into a view that does not adopt his metaphysical distinction between human animals and persons. The moral distinction he wishes to make is already incorporated into current legal theory. We make accommodations for mitigating circumstances and admit temporary insanity as a defense against legal or moral responsibility for an action. There is nothing in this practice that implies that there must be a standard of personal identity distinct from the standard for human beings. We are able to make the kind of moral judgments for which Locke thinks that we need the concept of personal identity without adopting his view. We can recognize that human moral agency can be undermined without treating the person as a distinct entity.

The issue of moral agency may cause another kind of problem for the view that there is nothing more to personal identity than the continued identity of a human being. We may be tempted to ascribe to nonhuman animals the qualities that make a typical human being a legitimate subject of human concern. Are all persons human beings? This question can productively be broken down into two quite different questions. One, "Are any nonhuman animals human beings?" is obviously answered in the negative. We can ask a more substantial question. "Are there other beings that are conscious, self-conscious, or deserving of our moral attention in the same way that a fully functional human being is?" It seems to be at least possible that there are some non-humans that have the relevant properties that we associate with typical adult human beings, deserving of some moral concern either based on their special relations to us or because of potentialities that they share with us and that we consider morally important. I see no principled reason for being certain that an artifact of some sort could not come to have either of these properties too.

Locke's example of waking Socrates and sleeping Socrates sharing a single human body brings to mind cases to reconcile with a theory of personal identity as the identity of a human being. We have come to associate a single human being with a more or less unified personality. Humans suffering from multiple personality disorder appear to have many distinguishable personalities within a single human body. In cases like these, the apparent presence of more than one consciousness in a single body raises doubts about the presumption that there is a one-to-one relation between human beings and personalities. One response to this deviation is to claim that there is a one-to-one relation between persons and personalities. If we choose this interpretation, there can be many persons connected to a single human being. This interpretation eliminates the possibility that personal identity and human identity could be the same thing, because a personality could cease to exist without the human being ceasing to exist. Indeed, one form of therapy for someone suffering from multiple personality disorder is to try to eliminate some of the personalities and combine others, to return the human patient to a less fragmented personality.

Clearly, something very unusual is occurring in cases of multiple personality. This phenomenon must somehow be accommodated within our conception of human identity. We are not obliged to create a new substantial kind to deal with these cases though. Instead, we may recognize that the presumption of psychological and dispositional unity in a human being is sometimes unjustified. Most human beings present a relatively unified personality, but that unity may be subject to dramatic fragmentation in response to psychological trauma. A human being may develop a disjointed or fragmented personality, the pieces of which are not smoothly coordinated. To me, this seems like a clear case of a damaged human being rather than the existence of more than one person. The treatment of these cases by trying to unify some of the personality fragments and expunge others lends support to the view that no substantial living thing is destroyed in the process.

6.3 CONCLUSIONS

If there are additional substantial kinds for human beings, it seems likely that they would deal with some of the psychological characteristics that I examined above. Properties such as being a conscious agent, having a unified personality, or being worthy of moral consideration are the best candidates. Nothing in the abstract metaphysical structure of substantial individuals and substantial kinds determines whether any of these characteristics mark a substantial change. The reality of a kind as a substantial kind depends on whether or not an entity of that kind ceases to exist if it loses the properties characteristic of that kind. Determining whether any of these properties define a substantial kind is an empirical as well as a conceptual matter. If any of these kinds are definitive of personal identity, then not all human beings are persons, some nonhuman animals may be persons, and there may be more than one person associated with a single human being.

I think that I have shown that we are not bound to accept the addition of a further substantial kind on the basis of any evidence that we have available to us. I am inclined to reject any of the possibilities as genuine natural kinds because it does not seem to me that something ceases to exist when it loses any of those properties. The loss of one or more of the faculties is a tragedy for the human being or nonhuman animal with the potential to achieve them. The loss of those properties will affect the way that we interact with that creature and that creature's ability to think about itself and other things, but it does not cause that creature to cease to exist. This judgment is subject to revision. Many of the cases that might strain the concept of a human being have never occurred, for example the case of a brain transplant. We have only a tenuous grasp on the nature of consciousness, one of the basic phenomena to be explained by a theory of personal identity. Our ability to explore the cognitive capacities of animals is imperfect. Much of the psychological evidence about personal identity is presented through the lens of preexisting philosophical biases. The empirical evidence relevant to personal identity is sketchy, and I have tried to emphasize throughout this work the empirical nature of the evidence that must go into our conception of the substantial kinds for living entities.

APPENDIX

Identity and Sortals

Why Relative Identity Is Self-Contradictory

In this appendix, I explore the role of substantial sortals in the relation of identity. The view I present here is an extension of my argument from the sections in Chapters 1 and 2 about substantial kinds. I have relegated this argument to an appendix because many readers may be willing to accept without argument the idea that identity is an absolute relation. It was important to include it though, because of a tendency to interpret my views from Chapter 3 as entailing that the same thing can be correctly identified as more than one kind of substantial individual at a time, even if those kinds have incompatible persistence conditions. Instead, I argue that the right way to think of these different kinds of individuality is to recognize that some of these individuals can overlap through composition without thereby being identical.

If we know that an object a is identical with an object b and that a is an F (where F is a substantial sortal), what role does that substantial kind play in the identity relation? I will consider three positions.

i. Identity is absolute, and neither sortal-dependent nor sortal-relative.
ii. Identity is absolute and sortal-dependent.
iii. Identity is sortal-relative.

In the end, I will argue for (i), that identity is absolute and neither sortal-dependent nor sortal-relative. The substantial sortal picks out the object about which one is making an assertion of identity. This is an important function because of the possibility of more than one thing occupying the same place at the same time. The sortal plays a referential role; it picks out what is referred to, rather than playing a role in the identity relation itself. The merits of this position are best seen in contrast with the other two options.

I think that the sortal just makes clear to which object we mean to refer, which is often ambiguous. In contrast, Wiggins thinks that the identity relation is always sortal-dependent. He has introduced a formal notation to

Appendix

accommodate this view. If a is a donkey and b is a donkey and a is identical with b, Wiggins would formulate the relation between a and b as follows:

$$\underset{\text{donkey}}{`a = b\text{'}}$$

which he takes to be a shorthand for 'a is a donkey, b is a donkey, and $a = b$.' He thinks that the fact that there is a substantial sortal under which any given living thing falls and the Identity of Indiscernibles, $(x)(y)[(x = y) \rightarrow (\phi)(\phi x \equiv \phi y)]$, are enough to prove that the identity relation will be "under a sortal." If a and b exist, then a must be some kind of thing and b must be some kind of thing. There must be a sortal predicate F which a satisfies and some predicate G which b satisfies. If $a = b$, then by the Indiscernibility of Identicals there must exist an F such that $a = b$ under F. These terms are used both to say what a thing is, "a is an F," and also to cover identity statements such as 'a is the same F as b.'

It follows from accepting his version of sortal-dependent identity and the Indiscernibility of Identicals that it is impossible that $a = b$ under sortal F, but $a \neq b$ under sortal term G, even if either a or b is a G. That is to say, that

$$R: \underset{f}{(a = b)} \& \underset{g}{(a \neq b)} \& (g(a) \vee g(b))$$

is false. Anyone who thinks that R is a real possibility can favor the relativity of identity at the expense of the Indiscernibility of Identicals, so Wiggins's argument against the relativity of identity depends on the acceptance of the Indiscernibility of Identicals. Because sortal-relative identity still appears to be a viable option, I will discuss it before returning to Wiggins's position.

Peter Geach and other supporters of relative identity argue that it is possible that a and b are identical when identified under one sortal concept but not identical when identified under another sortal concept.[1]

> I maintain that it makes no sense to judge whether things are 'the same,' or a thing remains 'the same,' unless we add or understand some general term – "the same F." That in accordance with which we judge whether identity holds I call a *criterion* of identity; this agrees with the etymology of "criterion." (Geach 1980, pp. 63–64)

In this quotation Geach has done little more than point out the importance of a sortal term in formulating an identity relation. It is interesting to note the extent of possible agreement between the three positions I am examining. All three can agree that every individual is an individual of at least one substantial kind. Differences do not crop up until the role of the sortal in the formulation of identity statements is addressed.

Geach's interpretation of the role of the sortal in the identity relation goes beyond what Wiggins can accept.

> Frege has clearly explained that the predication of "one endowed with wisdom" ("*ein Weiser*") does not split up into predications of "one" and "endowed with wisdom" ("*weise*"). It is surprising that Frege should on the contrary have constantly assumed that "x is the same A as y" does split up into "x is an A (and y is an A)" and "x is the same as (*ist dasselbe wie, ist gleich*) y." (Geach 1980, p. 176)

He considers it to be a real possibility that a and b are identical when identified under one sortal but nonidentical when identified under another sortal. Because he thinks this is possible, he thinks that the formulation of an identity statement such as '$a = b$' is incomplete because it makes no sense to assert the identity relation without specifying a sortal concept to which the identity relation is relative.

> It is as nonsensical to speak of identification apart from identifying some *kind* of thing, as to speak of counting apart from counting some kind of thing. A numerical word demands completion with a count noun; similarly for "the same" and "another." (Geach 1971, p. 289)

Geach has a counterexample that he thinks supports the relativity of identity because it causes problems for Wiggins's view that '$a = b$ under sortal F' is equivalent to 'a is an F, b is an F, and $a = b$.'

> Let us suppose that the recently ennobled Lord Newriche has been visiting the Heralds' College to consult the heralds about his coat of arms. The papers of his case are on the desk of Bluemantle; "Bluemantle" is a name *for* a herald, in official language, and is grammatically a proper noun. If Lord Newriche saw Bluemantle at the Heralds' College on Monday and Tuesday, then on Tuesday it would be true to say:

> (1) Lord Newriche discussed armorial bearings with some herald yesterday and discussed armorial bearings with the same herald again today. (Geach 1980, p. 174)

Or equivalently if we accept Wiggins's interpretation of identity:

> (3) For some x, x is a herald and Lord Newriche discussed armorial bearings with x yesterday and discussed armorial bearings with x again today.

Appendix

Geach thinks that we can analyze the statement

> (4) Lord Newriche discussed armorial bearings with some man yesterday and discussed armorial bearings with the same man today.

as equivalent to:

> (6) For some x, x is a man, and Lord Newriche discussed armorial bearings with x yesterday and discussed armorial bearings with x again today.

along the lines that Wiggins suggests. He also adds as an assumption (A) that whatever is a herald is a man.

> (A) For any x, if x is a herald, then x is a man.

If we use Wiggins's interpretation, (6) is true iff

> (7) Some A is a man, and Lord Newriche discussed armorial bearings with that (same) A yesterday, and Lord Newriche discussed armorial bearings with the same A today.

is true. If there is a change of personnel in the Heralds' College, Lord Newriche might have seen a different *man* on Monday and Tuesday but the same *herald*, namely Bluemantle.

> It is easy to see what has gone wrong.... (6) tells us that Lord Newriche discussed armorial bearings with something or other on two successive days, the same by some criterion or other, and this something-or-other *is* a man, whether tenselessly or omnitemporally. This does indeed follow from (2) or (3), and therefore from (1), by way of our additional premise: but it is a much weaker proposition than (4). "The same something-or-other, which is a man" does not boil down to "the same man." (Geach 1980, p. 176)

Geach thinks that he has pointed out the importance of specifying under which sortal, *man* or *herald*, we interpret the identity relation. Depending on which one we use, the statement will have different truth values, something that Wiggins argues is impossible. One problem with Geach's view is that *herald* is not really a substantial sortal. It is obvious that someone can survive the loss of heraldic office. Even if Geach is provisionally allowed to admit heralds to our ontology as spatiotemporally discontinuous objects, there remains a fundamental problem with his view.

E. J. Lowe has developed an argument that refutes the possibility of relative identity, and unlike Wiggins's attempt to do the same, it does not depend on the Indiscernibility of Identicals. His argument can be presented in the form of a reductio. Assume that it is possible for an individual to belong to distinct kinds with different criteria of identity. To continue on with Geach's case, we are presented with A, who is both a man and a herald. Something can cease to be a herald and still be a man. A different man at a different time can be the same herald. *Herald* and *man* are sortal kinds with different identity conditions. One can remain a man and cease to be a herald. What happens if A ceases to fulfill the criterion of identity associated with being a herald while continuing to fulfill the criterion for being a man? Does A now both exist and not exist? That is a contradiction. Or does A continue to exist as a man but cease to exist as a herald?

> The only way to evade the conclusion of this argument would be to attempt to relativize the very concept of *existence* itself to sortal distinctions – saying, for instance, that x might cease to exist *qua* ϕ but continue to exist *qua* ψ. (Lowe 1989, p. 57)

To explain this situation without contradiction, it is necessary to relativize existence, something Geach himself does not accept. Lowe has proven that relative identity is internally contradictory because it could only work if an entity is individuated by two distinct substantial sortals with distinct, and therefore potentially inconsistent, identity criteria. Two important maxims can be derived from Lowe's proof. When an individual belongs to two different substantial kinds, these kinds cannot have different criteria of identity. Also, "Where two sortal concepts are governed by different criteria of identity, it makes no sense to identify an individual falling under one of these concepts with an individual falling under the other" (Lowe 1989, p. 2). Wiggins also has a proof against the possibility of relative identity, but his is dependent on the Indiscernibility of Identicals (Wiggins 1980, pp. 19–20).

Wiggins examines and rejects a series of cases that seem to support R. The most relevant cases are ones in which the sortal F continues to apply to a and b, and sortal G seems to apply to a but not b. Examples include the relation between an object and the matter composing it. The correct way to deal with the putative cases of relative identity is to make a distinction between composition and identity. For example, in the case of a sexually produced zygote or single-celled organism the same matter composes both a functional individual and a genetic individual. Despite the complete material coincidence, these two kinds of individual are not identical because they have either different actual persistence conditions or at least potentially different persistence conditions. As paradoxical as it may sound, more than one thing can be in a place at a time. We must accept this because the disparity in identity conditions between the two kinds of individuals precludes their identity.

Appendix

Is there a difference between sortal-dependent identity as characterized by Wiggins and classical absolute identity as defined by Leibniz's Law? Wiggins thinks that they are equivalent. Why then does he emphasize the sortal's role to the extent that he does? Part of the answer is that he recognizes potential difficulties in determining which object among the possible overlapping objects in a place at a time one is referring to.

> There is no reason, however, to rule out *a priori* one other possibility: this is that the matter to be discovered in a certain place at a certain time may sometimes be usefully categorized at such widely different levels of explanation or theory that these categorizations are not commensurable with one another at all. There need be no common or agreed identification of what is there. This does not imply disagreements between identifications. It implies distinctness in the entities identified. Radically different kinds of thing may co-occupy some portion of space. (Wiggins 1980, p. 204)

When the covering sortal is explicit, this helps to avoid confusion that could arise given the number of things that could be in the same place at the same time. Other philosophers have recognized this fact.

> I think that Frege could agree with Geach that an utterance of the grammatical form "x and y are the same" might not have a clear truth-value, and that this situation might be remedied by adding a general term after the word "same." (Perry 1970)

'Sortal-dependent' seems to me to be a confusing way of referring to the proper role of the substance sortal in an identity relation. What are the characteristics of that relation in absolute classical identity?

Classical identity is defined by Leibniz's Law:

$$(x)(y)[(x = y) \equiv (\phi)(\phi x \equiv \phi y)]$$

Leibniz's Law is an attempt to define identity in terms of the complete congruence of predicates. It is a biconditional made up of the two principles, the Identity of Indiscernibles (or its converse, the Discernibility of Nonidenticals):

A. $(x)(y)[(\phi)(\phi x \equiv \phi y) \rightarrow (x = y)]$
B. $(x)(y)[\sim(x = y) \rightarrow \sim(\phi)(\phi x \equiv \phi y)]$

and the Indiscernibility of Identicals (or its converse, the Nonidentity of Discernibles):

C. $(x)(y)[(x = y) \rightarrow (\phi)(\phi x \equiv \phi y)]$
D. $(x)(y)[\sim(\phi)(\phi x \equiv \phi y) \rightarrow \sim(x = y)]$

Identity and Sortals

We have already had reason enough to believe that the Indiscernibility of Identicals is true. There is reason to believe, however, that Leibniz's Law is false because principle A (and therefore B) is false. I have been convinced by the counterexamples presented by Max Black.

> Isn't it logically possible that the universe should have contained nothing but two exactly similar spheres? We might suppose that each was made of chemically pure iron, had a diameter of one mile, that they had the same temperature, colour, and so on, and that nothing else existed? Then every quality and relational characteristic of the one would be a property of the other. Now, if what I am describing is logically possible, it is not impossible for two things to have all their properties in common. (Black 1952, p. 156)

Max Black has produced a tricky counterexample to the Identity of Indiscernibles. As he notes, any form of radially symmetric universe would function as a counterexample. It is a principle far more suspect than the Indiscernibility of Identicals. It is possible for there to be two distinct objects that do not differ from one another in either intrinsic or relational properties, making the Identity of Indiscernibles false. Because I do not accept one of the conditionals making up the definition of identity provided by Leibniz's Law, I reject it as a definition of the identity relation.

Because of my suspicion that the Identity of Indiscernibles is false, I make no use of it. However, nothing for which I argue depends on the denial of the principle, and I feel confident that there are no two real living things that do not differ in either intrinsic or relational properties. There is no better attempt to define identity than Leibniz's Law, so I will treat identity as an indefinable logical primitive.

We still have something to work from though, because we know that if x and y are identical then they must have all properties in common. The Indiscernibility of Identicals is true. If we know that $a = b$, we know that there is no difference between them. Any property that a has, b has. So if a is of kind F, then b is of kind F as well.

Identity is a primitive and absolute relation. Although using a sortal term may be the only way or the best way to clarify of what entity identity is being asserted, the sortal does not play any essential role in the formulation of that assertion of identity. It may give the *conditions* of identity and unity for an entity of that kind. Different sortals do not mark different *kinds* of identity. The sortal term will specify what kind of thing is asserted about which identity and give the conditions of identity.

Notes

1. BEYOND HORSES AND OAK TREES

1. See *Essay Concerning Human Understanding*, section 6, for Locke's discussion of the identity of human beings, which is similar to his treatment of the identity of other animals. This should not be confused with Locke's theory of personal identity in Book II.xxvii.9–29, which does not necessarily involve the continuation of a life.
2. Locke has a technical definition of 'body.' He uses the term to refer to a group of atoms or a mass of matter. A body, in this use of the term, continues to exist only so long as it is composed of just the atoms or matter that originally composed it. The matter may be rearranged without loss of identity.
3. Individuating horses and oak trees may not be as easy as it seems at first. Is the heartwood of an oak tree a part of the oak tree's continued life? The bark? Is a horse identical with the zygote from which it comes? Could that same zygote have resulted in twins? If so, which of the twins is identical with the horse we began with? Either, both, or neither? Is a transplanted heart valve a part of the horse? A *plastic* heart valve? I address questions like these later.
4. As a further deviation from the common run of cases, I will not use human beings as examples until the last chapter. The question of how to individuate people is complicated by controversial issues of personal identity. The only constraint that the questions of personal identity place upon my theory is that it be consistent with a plausible story, or more likely, a range of plausible stories regarding human identity.
5. See *Metaphysics*, Book VII (H), chapter 2, for an example of this. Here Aristotle uses the example of a house as a means of explaining the notion of a substance, though the primary substances he has in mind are organisms. For a more in-depth treatment of this issue in Aristotle's works, see Montgomery Furth's *Substance, Form, and Psyche: An Aristotelian Metaphysic,* section 19.iii (pp. 181–184).
6. For the history of early thought experiments see Nicholas Rescher's "Thought Experimentation in Presocratic Philosophy" and Peter King's "Mediaeval Thought-Experiments: The Metamethodology of Mediaeval Science," both in *Thought Experiments in Science and Philosophy.*
7. I discuss both root-heads and aphids in more depth in Chapter 5.
8. A notable exception to this rule is Kathleen Wilkes's *Real People.* Wilkes attempts to discuss the main issues of personal identity using only actual science and psychology as the source of her examples. She also attacks the legitimacy of thought experiments in her first chapter, which I cite later in this chapter.

9. I found the relevant facts for this story in Stephen J. Gould's *Ever Since Darwin*, pp. 171–178, and F. W. Kent's "The Size of Man."
10. My assumption does not entail the modal essentialism I argue for in Chapter 4. It does entail that there are some changes an entity cannot survive, which is enough to entail some sort of temporal essentialism, i.e., that for any entity there are some properties the loss of which it cannot survive.
11. " 'Tis true, there is ordinarily supposed a real Constitution of the sorts of Things; and 'tis past doubt, there must be some real Constitution, on which any Collection of simple *Ideas* co-existing, must depend. But it being evident, that Things are ranked under Names into sorts or *Species*, only as they agree to certain abstract *Ideas*, to which we have annexed those Names, the *Essence* of each *Genus*, or Sort, comes to nothing but that abstract *Idea*, which the general or *Sortal* (if I may have leave so to call it from *Sort*, as I do *General* from *Genus*,) Name stands for" (*Essay*, Book III.iii.15).
12. The formulation of (1b) leaves open the possibility that a thing could have been of a kind different from the kind it actually is, if it were of that kind from the start of its existence. I think that it is impossible for a thing to have been of a different substantial kind than it actually is because of issues involving the necessity of origin, but I do not argue for this view until Chapter 4.
13. Aristotle does not rule out the possibility of bare particulars in the *Categories*. Although he describes some kinds as substantial kinds, he leaves open the possibility that an individual could change substantial kinds without loss of numerical identity, i.e., without ceasing to exist. Furth (*Substance, Form, and Psyche: An Aristotelian Metaphysics*, p. 35) thinks that Aristotle omitted the metaphysical structure that would have eliminated this possibility because it would have forced him to give a coherent account of the complicated dependence of an atomic particular on its substantial form. Whether Furth is right about the reasons for this omission or not, this deviant interpretation of the *Categories* is possible.
14. I say that there is at least one because it is possible that there are more than one substantial kind that any given individual living thing instantiates. David Wiggins raises this possibility on p. 60 of *Sameness and Substance*. If a thing instantiates more than one substantial kind though, those kinds must have the same persistence conditions associated with them to prevent the possibility that the entity ceases to fulfill one of the sortals while continuing to fulfill the other.

2. THE BIOLOGICAL AND PHILOSOPHICAL ROOTS OF INDIVIDUALITY

1. See in particular C. M. Child's 1915 monograph, *Individuality in Organisms*.
2. I take no side in the dispute over the chronological ordering of the *Categories* and the *Metaphysics* or what Aristotle intended the relation between them to be.
3. Michael Ayers thinks that Locke fails to notice this distinction. See his *Locke: Epistemology and Ontology*, vol. 2, p. 219.
4. These examples are from Kripke, *Naming and Necessity*, pp. 115–116, notes 56 and 57.
5. Putnam has since distanced himself from the view I discuss and criticize here. See in particular Putnam, *Realism with a Human Face*, ch. 4, "Is Water Necessarily H_2O?"
6. Putnam, "The Meaning of 'Meaning,'" in *Philosophical Papers II*, p. 218.

7. This example is from Crawford Elder's "Higher and Lower Essential Natures."
8. The notion of a pattern I use in this section is derived from Daniel Dennett's article "Real Patterns," though the use I make of it is my own.
9. See the Appendix for a more complete explanation of overlapping individuals.

3. INDIVIDUALITY AND EQUIVOCATION

1. I limit my discussion to the *adult* higher animals in this section to avoid the ambiguities of individuation during the early stages of development when there is little, if any, integration among the cells that compose the embryo. I address this issue in section 5.2.
2. Not all particulars are material objects, e.g., sounds or thoughts are not material objects, but all living particulars are.
3. See van Inwagen, *Material Beings*, p. 287, note 21.
4. The distinction between causal interaction and causal integration is taken from Mishler and Brandon's "Individuality, Pluralism, and the Phylogenetic Species Concept."
5. Richard Dawkins defined the germ line as "that part of the bodies which is potentially immortal in the form of reproductive copies: the genetic contents of gametes and of cells that give rise to gametes" (*The Extended Phenotype*, p. 287). The germ-line cells are differentiated from the somatic cell lineages that form the body of an organism, and, at least during preformistic development, variation in the somatic cells is not heritable from one generation to the next unless it is passed on through the germ line.
6. Leo Buss, *The Evolution of Individuality*, p. 20. The details of the three forms of development are based on pp. 13–25 of this book.
7. Buss argues that this is at best an ideal that is only approximated in the higher animals and is not true at all for the majority of living things. I agree with Buss on this point and will return to it in later sections. For now though, because I am discussing one of the higher animals, I will use this idealization.
8. Wittgenstein discussed this notion in *Philosophical Investigations*: "And this is true. – Instead of producing something common to all that we call language, I am saying that these phenomena have no one thing in common which makes us use the same word for all, – but that they are related to one another in many different ways" (65); and "What is common to them all? – Don't say: 'There *must* be something common, or they would not be called "games" ' – but *look and see* whether there is anything common to all. – For if you look at them you will not see something that is common to *all*, but similarities, relationships, and a whole series of them at that" (66).
9. 'Superorganism' appears to have two rather different uses. One is to describe integrated living things that do not meet the stricter definition of 'organism.' The second use is to describe entities that are composed of groups of organisms and function as units of selection.
10. Dawkins cites Bonner's *On Development* as support for this distinction. Leo Buss's *The Evolution of Individuality* also emphasizes the importance of the single-celled bottleneck as well as access to the germ line as major forces for body plan innovations.
11. See section 2.4 and the end of the Appendix for a fuller account of how entities can overlap or even completely coincide.

4. THE NECESSITY OF BIOLOGICAL ORIGIN AND SUBSTANTIAL KINDS

1. *Man* and *centipede* are not really substantial kinds, but this is the example Mackie uses and we can use it constructively in this context.
2. Having the same origin implies having the same genotype. Having the same genotype, though a reliable indicator of having the same origin, does not logically imply it.
3. Some hermaphroditic animals can change from one sex to the other and back. See M. T. Ghiselin's *The Economy of Nature and the Evolution of Sex* (pp. 99–137) for further details.
4. Information about chromosomal variation among species is from Helena Curtis and N. Sue Barnes's *Biology*.

5. GENERATION AND CORRUPTION

1. Despite the urban myth to the contrary, Walt Disney was not frozen after death. He was cremated.
2. These components, or organelles, may descend from free-living prokaryotes that parasitically entered another prokaryote. I discuss this hypothesis later in this section.
3. See L. Margulis, *Symbiosis in Cell Evolution*. If her theory is correct, it lends support to the idea that symbiotic partners can become so integrated that they form a single functional unit.
4. This information on the root-heads comes from Stephen J. Gould, "Triumph of the Root-Heads."
5. I want to thank Michael Ghiselin for this example.

6. PERSONAL IDENTITY NATURALIZED

1. See section 5.3 for reasons why the functional individual is not coextensive with the developmental individual.
2. If it is discovered that this function is fulfilled by a biological system either more or less inclusive than the brain, it is that system to which I mean to refer in this section.
3. For a discussion of split-brain phenomena see Thomas Nagel's "Brain Bisection and the Unity of Consciousness"; for an alternative account see Kathleen Wilkes's *Real People*, ch. 5, "Being in Two Minds."

APPENDIX. IDENTITY AND SORTALS

1. See, for example, Nicholas Griffin's *Relative Identity*.

References

Ackrill, J. L., ed. 1987. *A New Aristotle Reader*. Princeton: Princeton University Press.
Ayers, Michael. 1981. "Locke versus Aristotle on Natural Kinds." *Journal of Philosophy* 78, pp. 247–272.
———. 1991. *Locke: Epistemology and Ontology*, vol. 2. London: Routledge.
Black, Max. 1952. "The Identity of Indiscernibles." *Mind* 61, pp. 1563–1564.
Bonner, J. T. 1974. *On Development*. Cambridge, Mass.: Harvard University Press.
———. 1988. *The Evolution of Complexity*. Princeton: Princeton University Press.
Boyle, James. 1996. *Shamans, Software, and Spleens. Law and the Construction of the Information Society*. Cambridge, Mass.: Harvard University Press.
Brandon, Robert, and Richard Burian, eds. 1984. *Genes, Organisms, Populations: Controversies over the Units of Selection*. Cambridge, Mass.: MIT Press.
Brasier, Clive. 1992. "A Champion Thallus." *Nature* 356, pp. 382–383.
Buss, Leo. 1987. *The Evolution of Individuality*. Princeton: Princeton University Press.
Child, C. M. 1915. *Individuality in Organisms*. Chicago: University of Chicago Press.
Curtis, Helen, and N. Sue Barnes. 1989. *Biology* (5th ed.). New York: Worth.
Dawkins, Richard. 1982. *The Extended Phenotype*. Oxford: W. H. Freeman.
Dennett, Daniel. 1991. "Real Patterns." *Journal of Philosophy* 88, pp. 27–51.
Descartes, René. 1984. *Meditations on First Philosophy*, trans. John Cottingham. *The Philosophical Writings of Descartes*. Cambridge University Press.
Dupré, John. 1993. *The Disorder of Things: Metaphysical Foundations of the Disunity of Science*. Cambridge, Mass.: Harvard University Press.
Ehrhardt, A., and H. Meyer-Bahlburg. 1981. "Effects of Prenatal Sex Hormones on Gender-Related Behavior." *Science* 211, pp. 1312–1318.
Elder, Crawford. 1994. "Higher and Lower Essential Natures." *American Philosophical Quarterly* 31, pp. 255–265.
Forbes, G. 1980. "Origin and Identity." *Philosophical Studies* 37, pp. 353–362.
Furth, Montgomery. 1988. *Substance, Form, and Psyche*. Cambridge University Press.
Gale, Richard. 1991. "On Some Pernicious Thought Experiments." In *Thought Experiments in Science and Philosophy*, Tamara Horowitz and Gerald Massey, eds. Lanham, Md.: Rowman and Littlefield.
Geach, P. T. 1971. "Ontological Relativity and Relative Identity." In *Identity and Individuation*, Milton K. Munitz, ed. New York: New York University Press.
———. 1980. *Reference and Generality* (3d ed.). Ithaca: Cornell University Press.

References

Ghiselin, M. T. 1974. *The Economy of Nature and the Evolution of Sex.* Berkeley: University of California Press.
— 1974. "A Radical Solution to the Species Problem." *Systematic Zoology* 23, pp. 536–544.
— 1987. "Species Concepts, Individuality, and Objectivity." *Biology and Philosophy* 2, pp. 127–143.
— 1988. "The Individuality Thesis, Essences, and Laws of Nature." *Biology and Philosophy* 3, pp. 467–471.
— 1989. "Sex and the Individuality of Species: A Reply to Mishler and Brandon." *Biology and Philosophy* 4, pp. 73–76.
— 1997. *Metaphysics and the Origin of Species.* Albany: State University of New York Press.
Gould, Stephen J. 1977. *Ever Since Darwin.* New York: W. W. Dalton.
— 1984. "A Most Ingenious Paradox." *Natural History* (December), pp. 20–28.
— 1992. "A Humungous Fungus Among Us." *Natural History* (July), pp. 10–16.
— 1996. "Triumph of the Root-Heads." *Natural History* (January), pp. 10–16.
Griffin, Nicholas. 1977. *Relative Identity.* Oxford: Clarendon Press.
Griffiths, Paul E., and Russel D. Gray. 1997. "Replicator II – Judgement Day." *Biology and Philosophy* 12, pp. 471–492.
Haeckel, E. 1879. *Das System der Medusen.* Jena: Gustav Fischer.
Harper, John. 1977. *Population Biology of Plants.* New York: Academic Press.
— 1985. "Modules, Branches, and the Capture of Resources." In *Population Biology of Plants*, Jackson, Buss, and Cook, eds. New Haven: Yale University Press.
Hull, D. 1965. "The Effect of Essentialism on Taxonomy: 2000 Years of Stasis." *British Journal of Philosophy of Science* 15, pp. 314–326; 16, pp. 1–18.
— 1976. "Are Species Really Individuals?" *Systematic Zoology* 25, pp. 174–191.
— 1978. "A Matter of Individuality." *Philosophy of Science* 45, pp. 335–360.
— 1980. "Individuality and Selection." *Annual Review of Systematics and Ecology* 11, pp. 311–332.
— 1989. "A Function for Actual Examples in the Philosophy of Science." In *What the Philosophy of Biology Is: Essays Dedicated to David Hull*, Michael Ruse, ed. Dordrecht: Kluwer.
Huxley, Julian. 1912. *The Individual in the Animal Kingdom.* Cambridge University Press.
Huxley, T. H. 1852. "Upon Animal Individuality." *Proceedings of the Royal Institute* (April 30), pp. 184–189.
Jablonka, Eva, and Marion J. Lamb. 1995. *Epigenetic Inheritance and Evolution: The Lamarkian Dimension.* Oxford: Oxford University Press.
Janzen, D. H. 1977. "What Are Dandelions and Aphids?" *American Naturalist* 111, pp. 586–589.
Jeon, Kwang. 1991. "Amoeba and x-Bacteria." In *Symbiosis as a Source of Evolutionary Innovation*, Margulis and Fester, eds. Cambridge, Mass.: MIT Press.
Kent, F. W. 1968. "The Size of Man." *American Scientist* 56, pp. 400–413.
Kinchin, Ian. 1994. *The Biology of Tardigrades.* Chapel Hill, N.C.: Portland Press.
King, Peter. 1991. "Mediaeval Thought-Experiments: The Metamethodology of Mediaeval Science." In *Thought Experiments in Science and Philosophy*, Tamara Horowitz and Gerald Massey, eds. Lanham, Md.: Rowman and Littlefield.

References

Kitcher, Philip. 1984. "Against the Monism of the Moment: A Reply to Elliot Sober." *Philosophy of Science* 51, pp. 616–630.
Kripke, Saul. 1972. *Naming and Necessity*. Cambridge, Mass.: Harvard University Press.
Lewontin, Richard. 1970. "The Units of Selection." *Annual Review of Systematics and Ecology* 1, pp. 1–18.
Locke, John. 1689. *An Essay Concerning Human Understanding* (2d ed.), P. Nidditch, ed. Oxford: Clarendon Press.
Lowe, E. J. 1989. *Kinds of Being: A Study of Individuation, Identity and the Logic of Sortal Terms*. Oxford: Basil Blackwell.
Mackie, Penelope. 1994. "Sortal Concepts and Essential Properties." *Philosophical Quarterly* 44, pp. 311–333.
Margulis, L. 1981. *Symbiosis in Cell Evolution*. San Francisco: W. H. Freeman.
Mayr, Ernst. 1975. *Evolution and the Diversity of Life*. Cambridge, Mass.: Harvard University Press.
McCann, Edwin. 1987. "Locke on Identity: Matter, Life, and Consciousness." *Archiv für Geschichte der Philosophie* 69, pp. 54–77.
McGinn, Colin. 1976. "On the Necessity of Origin." *Journal of Philosophy* 73, pp. 127–135.
McShea, Daniel. 1996. "Metazoan Complexity and Evolution: Is There a Trend?" *Evolution* 50, no. 2 (April), pp. 477–492.
Medawar, P. B. 1957. *The Uniqueness of the Individual*. London: Methuen.
Mellor, D. H. 1977. "Natural Kinds." *British Journal for the Philosophy of Science* 28, pp. 299–312.
Mishler, B. D., and R. Brandon. 1987. "Individuality, Pluralism, and the Phylogenetic Species Concept." *Biology and Philosophy* 2, pp. 397–414.
Mishler, B. D., and M. Donoghue. 1982. "Species Concepts: A Case for Pluralism." *Systematic Zoology* 31, pp. 491–503.
Nagel, Thomas. 1971. "Brain Bisection and the Unity of Consciousness." *Synthèse* 22, pp. 396–413.
Neilson-Jones, W. 1969. *Plant Chimeras* (2d ed.). London: Methuen.
Oinonen, E. 1967. "Sporal Regeneration of Bracken in Finland in Light of the Dimensions and Age of Its Clones." *Acta For. Fenn.* 83, pp. 3–96.
Perry, John. 1970. "The Same F." *Philosophical Review* 79, pp. 181–200.
Price, Marjorie. 1977. "Identity Through Time." *Journal of Philosophy* 74, pp. 201–217.
Prior, Arthur. 1960. "Identifiable Individuals." *Review of Metaphysics* 13, pp. 684–696.
Putnam, Hilary. 1975. *Philosophical Papers II: Mind, Language, and Reality*. Cambridge University Press.
———. 1990. *Realism with a Human Face*. Cambridge, Mass.: Harvard University Press.
Quine, W. V. 1950. "Identity, Ostention, and Hypostasis." *Journal of Philosophy* 47, pp. 621–632.
Rescher, Nicholas. 1991. "Thought Experimentation in Presocratic Philosophy." In *Thought Experiments in Science and Philosophy*, Tamara Horowitz and Gerald Massey, eds. Lanham, Md.: Rowman and Littlefield.
Salmon, Wesley. 1984. *Scientific Explanation and the Causal Structure of the World*. Princeton: Princeton University Press.

References

Sanford, David H. 1993. "The Problem of the Many, Many Composition Questions, and Naive Mereology." *Noûs* 27, pp. 219–228.
Simon, H. A. 1962. "The Architecture of Complexity." *Proceedings of the American Philosophical Society* 106, pp. 467–482.
Smith, Myron, Johann Bruhn, and James Anderson. 1992. "The Fungus *Armillaria bulbosa* Is among the Largest and Oldest Living Organisms." *Nature* 356, pp. 428–431.
Sterelny, Kim, Kelly C. Smith, and Michael Dickison. 1996. "The Extended Replicator." *Biology and Philosophy* 11, pp. 377–403.
Strawson, P. F. 1959. *Individuals*. London: Methuen.
Svitil, Kathy. 1993. "Fungus Among Us." *Discover* (January), pp. 69–70.
van Inwagen, Peter. 1990. *Material Beings*. Ithaca: Cornell University Press.
    1993. "Naive Mereology, Admissible Valuations, and Other Matters." *Noûs* 27, pp. 229–234.
Wiggins, David. 1967. *Identity and Spatio-Temporal Continuity*. Oxford: Oxford University Press.
    1980. *Sameness and Substance*. Cambridge, Mass.: Harvard University Press.
Wilkes, Kathleen. 1988. *Real People*. Oxford: Clarendon Press.
Williams, M. B. 1985. "Species Are Individuals: Theoretical Foundations for the Claim." *Philosophy of Science* 52, pp. 578–590.
    1992. "Species: Current Uses." In *Keywords in Evolutionary Biology*, Keller and Lloyd, eds. Cambridge, Mass.: Harvard University Press.
Wilson, D. S. 1997. "Altruism and Organism: Disentangling the Themes of Multilevel Selection Theory." *American Naturalist* 150 (Suppl.), pp. S122–S134.
Wilson, Edward O. 1975. *Sociobiology*. Cambridge, Mass.: Belknap Press of Harvard University Press.
Wittgenstein, Ludwig. 1958. *Philosophical Investigations,* G. E. M. Anscombe, trans. New York: Macmillan.
Zemach, Eddy. 1976. "Putnam's Theory on the Reference of Substance Terms." *Journal of Philosophy* 73, pp. 116–127.

Index

akratic states, 112
animals, higher, 48, 89
  behavioral integration of, 52
  common properties of, 9, 56
  development of, 99
  as functional individuals, 91
  as paradigmatic individuals, 9, 58
aphids, 11, 88, 99
archetypes, Platonic, 17
Aristotle
  explanation of biological form, 28–29
  on substantial forms, 17, 29
  use of artifacts to explain the individuation of living things, 6
Armillaria bulbosa, 23–25, 59, 89
asexual reproduction, *see* reproduction, asexual

biological origins
  and continuity with zygote, 78
  and *d*-continuity, 78–79, 90
  necessity of, 37, 72, 79
  and sexual phenotype, 80–83
Black, Max, 125
Bonner, J. T., 92
Buss, Leo, 93–96

cancer, 93
compossibility, 73
concepts as tools, 15

Darwin, Charles, 17
Dawkins, Richard, 26, 64–66
  definition of an organism, 52
  on selfish genes, 67
Descartes, René, 12, 114
development
  "bottlenecks," 26, 65–66, 100–101
  environmental factors in, 80–81
  of the functional individual, 91–92

and multicellularity, 96
patterns of, 54, 94
Disney, Walt, 86, 130

Empedocles, 28–29
essential properties
  assumptions for, 16
  and biological species, 16
  as derived from the causal theory of reference, 37
  nominal, for Locke, 30
  sortal, challenges to, 70
  temporal, 16, 70–71
evolutionary individual, 25

folk ontology
  and natural kinds, 57
  role of in conceptual analysis, 9
Forbes, Graeme, 75–77
functional individuality, 89–99
  evolution of multicellularity, 92, 93
  and human beings, 106
  multicellular functional integration, 87–89
  and organ transplant, 107
  persistence conditions for, 99
  in plants, 97
  symbiosis and, 98
  zygote as, 77–78, 90
functional integration, 63–64

Gale, Richard, 15
Geach, Peter, 120–123
genet, 26, 88
genetic individuality, 64–66
  and human beings, 105
  possible forms of growth, 86–89
Ghiselin, Michael, 61–62, 130

Haldane, J. B. S., 11
Harper, John, 26, 65–66, 88

135

Index

hermaphrodites, 80
Hull, David, 10, 67
Huxley, J. S., 57, 58, 64
Huxley, T. H., 25–26, 48, 64

Identity of Indiscernibles, *see* Leibniz's Law
Indiscernibility of Identicals, *see* Leibniz's Law

Janzen, Daniel, 25, 88

Kripke, Saul, 36–39

Leibniz's Law, 120, 124–125
Lewontin, R. C., 66–67
Locke, John
   criteria for the continuation of life, 2–3, 6
   criticism of Aristotle's use of substantial forms, 31
   definition of 'sortal', 17
   material constitution of living entities, 2
   mechanistic account of individuation, 31, 103
   nominalist account of individuation, 29–33
   on the role of personal identity in moral judgments, 115–117
Lowe, E. J., 19, 122–123

Mackie, Penelope, 69
Margulis, Lynn, 63
McGinn, Colin, 77–79
metamorphosis, 7–8, 43, 100
metazoan, *see* animals, higher
Modern Synthesis, 94–95
monophylogeny, 84
multiple personality disorder, 117

Nanomia cara, 6–7
natural kinds, 20
   empirical revision of, 44, 46
   as patterns, 42
   Putnam's account of, 36, 39–42
   subjectivity in recognition of, 45

ontological pluralism, 46–47, 59, 68
origins, *see* biological origins

Parfit, Derek, 14
particulars, 49, 60–62
   bare, 18
   non-material, 129
   substantial, 18

Penis-at-Twelve, 82
personal identity
   the brain's role in, 108–109, 112
   as the continuation of a human functional individual, 111–118
   and memory, 109–111, 112
pluralism, *see* ontological pluralism

ramet, 26
reproduction, asexual
   as analogous to growth, 65, 88
   in plants, 8
resurrection, 101–104
rhizocephalans, 2, 11, 98–99, 100–101

Ship of Theseus, 12, 108
siphonophores
   comparison with jellyfish, 7
   development of, 6–7
slime mold, life cycle of, 8, 97
sortal
   definition of, 17
   empirical revision of, 34
   phase, 17
   substantial, 17, 18
sortal-relative identity, 120–122
species, biological
   essentialism and, 16, 17
   as individuals, 83
   Linneaus's views on, 83
   numerical phenetics, 83
   as substantial kinds, 29, 34
Strawson, P. F., 60
substantial kinds
   empirical nature of, 20
   modal attributes of, 36
   person as a, 113–115
   role in the identity relation, 119
superorganism, 129
symbiosis
   and the evolution of eukaryotic cells, 63, 98
   in lichens, 9

tardigrades, 86, 101–102
thought experiments
   in discussions of identity, 12
   history of in philosophy, 9
   lack of background conditions for, 10–11
transubstantiation, 20–21
transworld identity, 73, 77

unit of selection, 54, 67–68

Index

van Inwagen, Peter
  on composition, 51
  on imaginary examples, 13
  individuation of a life, 4
  on interrupted lives, 102–103
  view of personal identity, 109

Wiggins, David
  defense of sortal essentialism, 70
  rejection of sortal-relative identity, 123–125
  on the role of sortals in identity relations, 17, 119–120
  on substantial kinds as sortals, 33–35

Wilkes, Kathleen, 13, 14, 127
Wilson, E. O., 7, 63, 130
Wittgenstein, Ludwig, 1, 129

xenotransplantation, 97–98

Lightning Source UK Ltd.
Milton Keynes UK
UKHW03f0117260318
320034UK00001B/60/P